TOURISM, DEVELOPMENT AND TERRORISM IN BALI

Voices in Development Management

Series Editor:
Margaret Grieco
Napier University, Scotland

The Voices in Development Management series provides a forum in which grass roots organisations and development practitioners can voice their views and present their perspectives along with the conventional development experts. Many of the volumes in the series will contain explicit debates between various voices in development and permit the suite of neglected development issues such as gender and transport or the microcredit needs of low income communities to receive appropriate public and professional attention.

Also in the series

Women Miners in Developing Countries
Pit Women and Others
Edited by Kuntala Lahiri-Dutt and Martha Macintyre
ISBN 0 7546 4650 5

Africa's Development in the Twenty-first Century
Pertinent Socio-Economic and Development Issues
Edited by Kwadwo Konadu-Agyemang and Kwamina Panford
ISBN 0 7546 4478 2

Accountability of the International Monetary Fund
Edited by Barry Carin and Angela Wood
ISBN 0 7546 4523 1

Social Exclusion and the Remaking of Social Networks
Robert Strathdee
ISBN 0 7546 3815 4

Design and Determination
The Role of Information Technology in Redressing Regional Inequities in the Development Process
Stephen E. Little
ISBN 0546 1099 3

Tourism, Development and Terrorism in Bali

MICHAEL HITCHCOCK
London Metropolitan University, UK

I NYOMAN DARMA PUTRA
The University of Queensland, Australia

ASHGATE

Published by
Ashgate Publishing Limited
Gower House
Croft Road
Aldershot
Hampshire GU11 3HR
England

Ashgate Publishing Company
Suite 420
101 Cherry Street
Burlington, VT 05401-4405
USA

Ashgate website: http://www.ashgate.com

British Library Cataloguing in Publication Data
Hitchcock, Michael
Tourism, development and terrorism in Bali. - (Voices in development management)
1. Tourism - Indonesia - Bali Island 2. Bali Bombings, Kuta, Bali, Indonesia, 2002 - Economic aspects 3. Terrorism - Indonesia - Bali Island - Economic aspects 4. Bali Island (Indonesia) - Economic conditions
I. Title II. Putra, I Nyoman Darma
338.4'7915986

Library of Congress Cataloging-in-Publication Data
Hitchcock, Michael.
Tourism, development and terrorism in Bali / by Michael Hitchcock and I Nyoman Darma Putra.
p. cm. -- (Voices in development management)
Includes bibliographical references and index.
ISBN-13: 978-0-7546-4866-6
1. Tourism--Indonesia--Bali Island. 2. Terrorism--Indonesia--Bali Island. 3. Bali Bombings, Kuta, Bali, Indonesia, 2002. 4. Globalization--Economic aspects--Indonesia--Bali Island. I. Putra, I Nyoman Darma. II. Title.

G155.I5H57 2007
338.4'79159862--dc22

2006032965

ISBN: 978-0-7546-4866-6

Printed and bound in Great Britain by MPG Books Ltd, Bodmin, Cornwall.

Contents

List of Tables

List of Figures

All photographs by the authors unless otherwise stated

About the Authors

Professor Michael Hitchcock is Director of the International Institute for Culture, Tourism and Development at London Metropolitan University. He has a D.Phil from Oxford University (Linacre College 1983) based on fieldwork in Indonesia and a BA Hons (1978) from Queen's University, Belfast. Before taking up his current post, he taught Development Sociology at Hull University and was a member of the Centre for Southeast Asian Studies. He has written and edited 11 books and is the author of over 80 papers, many in refereed journals.

I Nyoman Darma Putra has a PhD from the University of Queensland and is a lecturer in the Indonesian Department, Faculty of Letters at Udayana University. He has written extensively on tourism, media, and literature in Indonesian, and has edited one book in English on Bali and two others in Indonesian. He has recently had three papers in English accepted on the Bali Bombings for internationally refereed journals. He is a Postdoctoral Research Fellow at the University of Queensland.

Acknowledgements

The authors are very grateful to the British Academy, the British Council, the Sutasoma Trust and the ASEAN-EU University Network Programme (Europe Aid) for their generous support with this research. We would like to thank Prof Richard Butler and Prof David Harrison for their helpful comments on earlier drafts of chapters in this book and to Nick Blackbeard for sharing his insights on the Bali bombings crisis. Our thanks are due to Udayana University and London Metropolitan University for their continuing support, and we are especially grateful to the *Bali Human Ecology Studies Group* (Bali HESG). In particular we would like to thank the late Prof Dr I Gusti Ngurah Bagus and Prof Adnyana Manuaba for their wholehearted encouragement. Thanks are also due to Prof Wayan Ardika, Dean of the Faculty of Letters, Udayana University, and Dr AAPA Suryawan Wiranatha, M.Sc., the Head of Centre Research for Cultural and Tourism, Udayana University, for their administrative support and discussion.

The idea for this book was developed from five papers, three of which were co-authored, that were published between 2001 and 2006. This book draws upon the research undertaken for these publications and we are grateful to the editors of the journals concerned for publishing the following papers:

Hitchcock, M. (2001) 'Tourism and total crisis in Indonesia: the case of Bali', *Asia Pacific Business Review* (Winter) 8:2, 101–120.

Hitchcock, M. (2004) 'Margaret Mead and tourism: anthropological heritage in the aftermath of the Bali bombings', *Anthropology Today* (June) 20:3, 9–14.

I Nyoman Darma Putra and M. Hitchcock (2005) 'Pura Besakih: a world heritage site contested', *Indonesia and the Malay World*, 33:96, 225–237.

I Nyoman Darma Putra and M. Hitchcock (2005) 'The Bali bombings: tourism crisis management and conflict avoidance', *Current Issues in Tourism*, 8:1, 2005, 62–76.

I Nyoman Darma Putra and M. Hitchcock (2006) 'The Bali bombs and the tourism development life cycle', *Progress in Development Studies*, 6:2, 157–166.

Abbreviations

ABC	Australian Broadcasting Corporation
ABRI	*Angkatan Bersenjata Republik Indonesia* [Armed Forces of the Republic of Indonesia]
ASEAN	Association of South-East Asian Nations
ASEAN-EU	Association of South-East Asian Nations-European Union
BALI HESG	Bali Human Ecology Studies Group
BTA	Bali Tourism Authority
JI	Jemaah Islamiyah
KPM	Royal Packet Navigation Company
MUI	*Majelis Ulama Indonesia* [Indonesian Islamic Council]
NGO	Non Governmental Organization
NV	*Naamloze Vennootschap* [Inc., Ltd]
PATA	Pacific Asia Tourism Association
PDIP	*Partai Demokrasi Indonesia Perjuangan* [Indonesian Democratic Party for Struggle]
SARS	Severe Acute Respiratory Syndrome
SCETO	Sociètè Centrale pour l'Equipment Touristique Outre-Mer
TALC	Tourism Area Life Cycle
TNCs	Trans National Corporations
UK	United Kingdom
UNESCO	United Nations Educational, Scientific and Cultural Organization
USA	United States of America

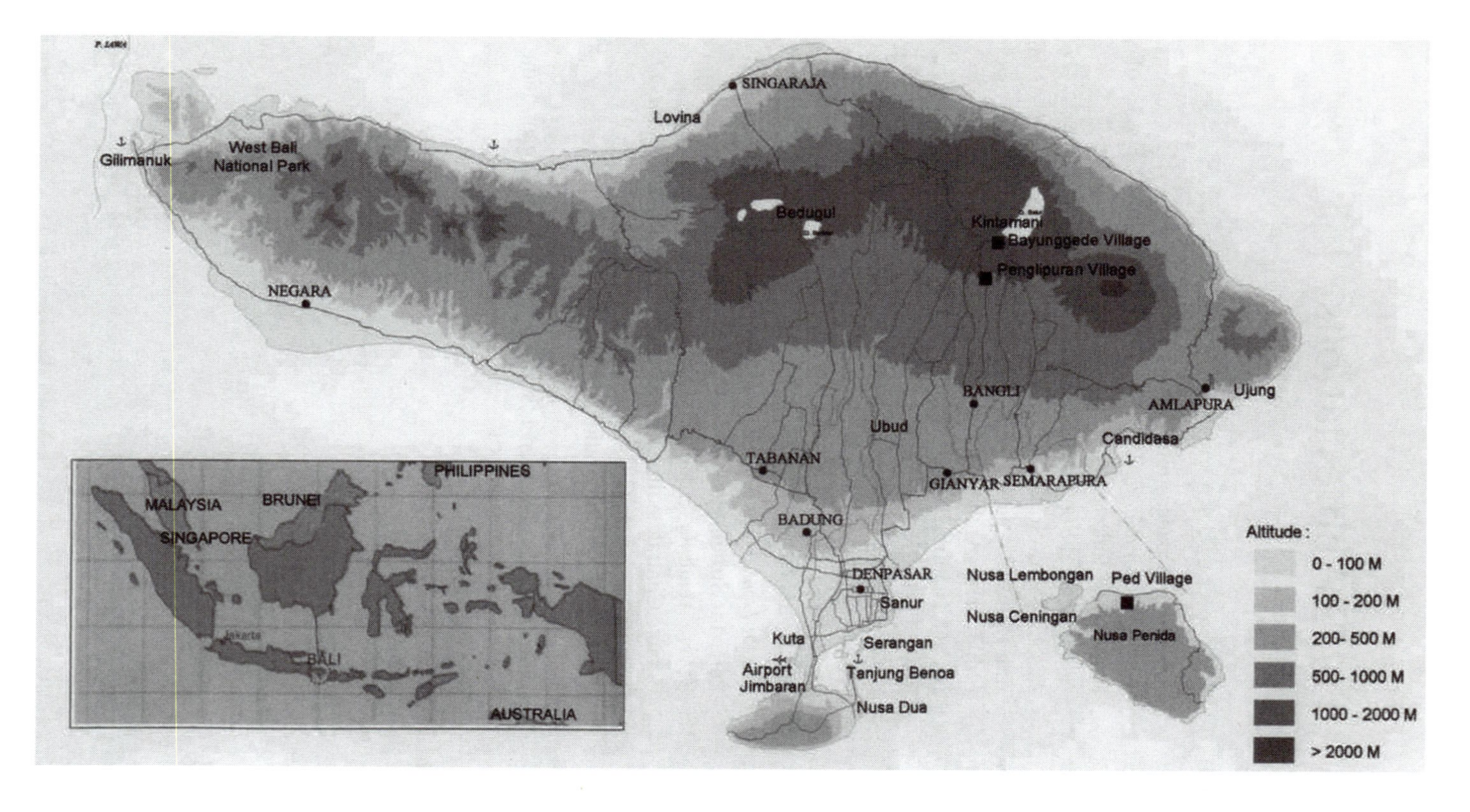

Figure 1.1 A map of Bali
Source: Ketut Sardian

Chapter 1

Introduction: A Paradise Globalized

Figure 1.2 Cartoons dealing with development and globalization are popular in Bali. This one is by Jango Pramartha

Bali is a small, but very renowned island situated eight degrees below the equator between Java and Lombok. What is less well known, even by some tourists who frequent the island, is that Bali is part of the vast Republic of Indonesia, the world's fourth largest nation in terms of population, a country that stretches five thousand kilometres from Sumatra to central New Guinea and comprises fourteen thousand islands. If these statistics are not impressive enough, one should also bare in mind that Indonesia is the world's largest Islamic country, though it is, in accordance with the national code of *Pancasila*, a multifaith state, comprising Muslims, Christians, Buddhists and Hindus. Pancasila means 'five principles' and the first principle, represented by a star on the national coat of arms, represents a belief in God and in practice this applies to any one of the country's formally recognised religions.

Bali is home to the republic's largest population of Hindus who comprise around 92% of the island's population, making it something of an anomaly within the wider context of Indonesia. Despite its distinctiveness, Bali occupies less than 0.3% of the land surface of Indonesia and is only 200 kilometres from east to west and 100 km from north to south. Bali's profile is, however, disproportionate to its size and it is without doubt better known worldwide than the nation in which is located, Indonesia.

Bali's fame is due to its renown as a tourist paradise, a reputation that has been reinforced by being the world's top island destination for four years in a row by readers of the American magazine *Travel and Leisure* (*The Bali Times* 15–21 July, 2005). Viewed from the perspective of marketing, 'Bali' is an easily recognised, remarkably successful and enduring brand that adds lustre to a wide variety of goods and services that take its name from 'Bali-style' hotels, to restaurants foodstuffs, beverages, works of art and fashion goods. Such is its lustre that other destinations in Indonesia, notably the island of Lombok, have to be marketed in terms of their better-known neighbour as 'the new Bali'. Travel agents in Indonesia actively promoted other Indonesian destinations by packaging them as 'Bali and beyond'. This moniker is often applied to destinations that lie beyond the borders of Indonesia, as if the reputation of each new island paradise rests on its similarity to the original, the Indonesian island of Bali. Even Dr Mahathir Mohamad, the former prime minister of Malaysia, invoked the name 'Bali' when commenting on future developments in Langkawi, arguing that the island could be as world famous as Bali if more conscientious efforts were made to put it on the global map (*Northern Region News*, April 3, 2006).

The name 'Bali' may have become something of an all purpose brand worldwide, but one should not lose sight of the fact that most products marketed in this fashion have some association, no matter how tenuous, with the industry with which the island is intimately associated, tourism. So pervasive is the notion that Bali is a 'tourist paradise' that it is not unreasonable to argue that the island is one of a number of global tourist brands, a fact that can be confirmed by looking at the sheer diversity of the countries of origins of the visitors who choose to holiday there, a veritable 'paradise globalized'.

Of course, the terms 'globalization' and 'paradise' are very vague, the former having widespread currency in contemporary social thought and the latter being one of the most clichéd marketing tools of tourism. Both terms moreover have a whiff of millenarianism about them, the former heralding a world without borders and the demise of the nation state, the latter representing the good life after death for three of the leading world religions, Judaism, Christianity, and Islam. Similar notions are encountered in other religions, notably the Hindu-Balinese concept of *suarga*, a place where the dead finally reside when they are released from the cycle of birth, death and rebirth. In accordance with Hindu teachings, only those who live above reproach are freed from the trauma of rebirth. Vague and even mythical that these terms may be, they have particular analytical value when applied to Bali's contemporary situation. Simply put, Bali would not be such a popular destination if it were not widely perceived as a paradise, and certain economic and political benefits would not have accrued to the island had it not been welcoming to foreign interests, notably global trade. Bali's success, however, in attracting both tourists and investments from around the world does not come without problems and as this book documents, the islanders have had to be incredibly resourceful in coping with the downside of being a global paradise, not least because the world has been very turbulent in recent times, notably in the Asia Pacific region.

A global industry such as tourism may be exposed to the vicissitudes of political and economic instability, but it is important to note that as Bali's leading industry tourism brings undoubted economic benefits, making the island one of Indonesia's wealthiest provinces. It is not tourism alone that creates this economic wellbeing, but the economic impact of a whole host of service sector (e.g. banking, retail, education) and small scale manufacturing industries (e.g. handicrafts, garments). The developmental impacts of tourism and related industries is not limited to Bali, but extends well beyond the island to Central and East Java, and the islands of eastern

Indonesia, notably Lombok. Goods produced in these areas are exported worldwide via Bali, using the infrastructure created by tourism, and Bali itself has become a much sought after source of skilled labour in hospitality in Indonesia and beyond. Such is the importance of tourism that former stalwarts of the Balinese economy such as agriculture are in relative decline, leading to an even greater dependence on tourism as a global phenomenon.

It is widely agreed that what makes Bali so attractive to foreign and domestic visitors is its combination of an equitable climate, unique civilisation and outstanding natural beauty. In particular visitor surveys draw attention to the hospitality and friendliness of the Balinese people. There is however, a substantial literature decrying the alleged threat posed by tourism to Bali's indigenous culture, much of it appearing towards the latter end of the 20th century. This book, however, describes how the Balinese have used and adapted their cultural resources as 'cultural strategists' or as a 'cultural solution' (Chapter 12) to weather a series of crises that have beset the island since the onset of the Asian Crisis in 1997, most notably the Bali Bombings of 12 October 2002, and the later bombings of 2005. Crises of this severity are not new in Bali, the island having for example suffered markedly during the Great Depression of the 1930s, Second World War and the turbulence of 1965, but these events occurred before the island had developed and become dependent on a major tourism industry. The crises that have struck Bali in more recent times have not only been economic and political, but have also been cultural, namely the rejection of a government-backed proposal to establish a World Heritage Site at the Besakih mother temple (Chapter 7).

In order to understand how Bali copes with the ups and downs of being part of the global economy, we examine the origins and growth of tourism (Chapter 2) and how Bali became an internationally recognised brand (Chapter 3), as well as the ramifications of a number of recent crises – the Asian Crisis (Chapter 8), the World Heritage Site débâcle of Besakih and the Bali Bombings (Chapters 9, 10, 11) – with a view to assessing the underlying causes of the island's resilience and the sustainability of tourism. In particular we look at how local cultural strategies were adopted to minimize the impact of the most dramatic recent crisis, the bombings, and restore confidence among the local and international business community. We also analyse the role played by local entrepreneurs in tourism and how they manage to compete in a global market (Chapter 5). The book also examines the highly variable impact of tourism economically on one of Bali's most important residential units, the traditional village, by looking at separate case studies (Chapter 4). Furthermore, the book investigates Balinese encounters with others, notably Westerners, and how these have been represented in popular literature (Chapter 6), as well as the local discourse on how the islanders might cope with being drawn into the global economy (Chapter 12).

The Globalization of Bali

If globalization involves the exchange and flow of economic and intellectual items including goods, knowledge, values and images, as well as people, on a worldwide scale, then there are a number of potential start dates for this process in Bali. If we were to follow Wallerstein (1976) and Wolf (1982) and argue that globalization is historically rooted in the expansion of trans Atlantic trade in the 16th century and the opening up of European oriented worldwide networks, then its onset in Bali could be pinpointed to 1597. The occasion was the arrival of a Dutch ship captained by Cornelius de Houtman, though he was not the first European to have contact with the island (Picard 1996, 18). Not only were goods and ideas (about cartography) exchanged, but the encounter left an enduring impression on Western consciousness, partly because of the men they left behind. Emanuel Roodenburg, a sailor from Amsterdam went ashore as a messenger, but did not return; Jacob Claaszoon, a seaman from Delft, also jumped ship just as the expedition set sail back to Holland (Hanna 1976, 10). Both seem to have entered the service of the Dewa Agung, perhaps not of their own volition or because they were unwilling to face the rigours of a sea voyage back to Europe. Whatever the reason they both took Balinese wives and learned the local language, and when the next Dutch expedition arrived in 1601, Roodenburg was able to act as a translator and general informant and advisor (Hanna 1976, 12). Early travel writers made much of these encounters in the years that followed, often illustrating their accounts with fanciful depictions of Balinese customs that were culled from various – and not necessarily Balinese – sources.

Alternatively, if we were to follow Giddens (1991) and to argue that the first truly global experience belongs to the 20th century and then we would link the onset of global social relations in Bali with the period of 'high colonialism', roughly 1870 to 1940 (Scholte 1997, 24), and in particular to the events of 1906 and 1908. This occasion saw the arrival of colonial troops belonging to the Netherlands East Indies to settle a dispute over a wrecked Chinese ship that had sailed under a Dutch flag; it ended with the annexation of Bali's last remaining independent kingdoms. The soldiers landed on the beach of Sanur in September 1906, now one of the island's best-known tourist resorts, and were marching towards what is now the capital of the island, Denpasar, when they encountered the royal household of Badung dressed in white. The king and his retinue had chosen to fight to the death rather than accept foreign domination in a suicidal battle known as the *puputan*, which is today commemorated as one Indonesia's struggles for self-determination. Since the Dutch were armed with the latest European military technology there was no question as to the outcome, and the Balinese were mowed down in a bloody and profoundly unequal battle. A second massacre took place in Klungkung in 1908. Preparations to develop tourism began a few weeks before the demise of the last independent Balinese kingdom; the publication of this book coincides with the hundredth anniversary of these grisly events.

Given that globalization is characterized by the intensification of social relations across the world, linking distant localities so that experiences in one location are influenced by events taking place far away in another and vice versa (Giddens 1991, 64), then the Balinese *puputans* are highly significant. There were journalists at hand to witness these events and when the massacres were reported back in Europe Holland's image as a responsible colonial power was severely tarnished. Strongly criticised back home for the killing of civilians, the colonial regime introduced more ethical policies such as the building of schools and the introduction of measures to help preserve local culture (Picard 1997). As part of this policy – and partly to atone for the massacres – the development of tourism in Bali became a priority for the government of the Netherlands East Indies. This was also a realistic strategy since the island lacked lands suitable for colonial plantations and produced few export commodities (Boon 1977). The origins of Bali's tourism industry can thus be traced back to 1908, a few weeks before the last Balinese kingdom fell, when representatives of commercial banks, railroads, insurance companies and shipping agencies founded in Batavia (now Jakarta) the Association of Tourist Traffic in Netherlands India (Picard 1996, 23).

As has been noted by Cheater, tourism is part of the process of globalization with '...an immense potential for affecting global relations, socio-political as well as financial and ecological' (1995, 125). In a similar vain, Lanfant has argued that the pressure of tourism causes societies to progressively open themselves to the world economy and to conceive of their '...dynamism and place in the world according to allogenous criteria' (1995, 5). This is also certainly true of Bali where the onset of globalization and subsequent developments have been so closely connected with tourism that it is almost impossible to talk about one without the other. Tourism is not of course the only globalizing force within the island, but it is certainly one of the most pervasive, not least when one considers the role of the media. Take for example the coverage of the cremation of the late Niang Muter (1910–2004), the twin sister of the last king of Ubud, Cokorda Gde Agung Sukawati, on 22 July 2004 whose elaborate funeral pyre was reported on television and radio worldwide, notably in Britain and Australia. The event was not only linked to the early 20th century, when Niang's long life began, but was connected to contemporary Australia through the marriage of her stepson to an Australian woman. Niang's grand daughter, who is half Australian, was featured in the ABC's Foreign Correspondent's coverage of the event in an attempt to connect with Australian audiences; it doubtless also helped reassure Australians, nervous about returning to Bali in the aftermath of the bombings, about the benign side of Bali. The presence of foreigners at Niang's cremation harked back to an earlier funeral, that of her brother in 1978. Cokorda's friend, Rudolf Bonnet, the painter who set up Pita Maha with Walter Spies and a group of Balinese artists in the 1930s and founded the Puri Lukisan Museum, was cremated alongside him. The cremation was thus also linked to Bonnet's family in the Netherlands and received widespread media coverage. Had the funeral taken place in another former kingdom on another Indonesian island and not in the well-known destination of Ubud on the very well known island of Bali it seems likely that the worldwide media would not have shown so much interest. As it happened, large numbers of tourists turned out to witness the spectacle, imbuing the event with a truly cosmopolitan feel, a fitting end perhaps to a member of a family who had done so much to put the relatively small kingdom of Ubud on the international map.

In view of the widespread publicity surrounding this ceremony it would perhaps be tempting to mark this as the occasion when Bali's long march that began in 1597 or 1906, depending on one's point of view, towards global incorporation finally achieved fruition. There are, however, good reasons for seeing this event as belonging to a continuum that begins with the fall of President Suharto, who was closely associated with Bali's tourism development, in 1998, and leads up to the dramatic downturns in tourism arrivals of the 21st century caused by the bombings of 2002 and 2005. This book does not attempt to cover the origins of tourism's fascination with Bali and its dramatic and controversial rise during the Suharto period, since there are two admirable books in particular that cover these issues: Adrian Vickers's *Bali: A Paradise Created* (1989) and Michel Picard's *Bali: Cultural Tourism and Touristic Culture* (1996). This book pays homage to Vickers and takes off where Picard ends with the decline of Suharto's regime and the onset of the Asian Crisis of 1997–98, the largest stock market crash since the Great Depression. It charts the turbulence that has afflicted the island at a time of market uncertainty and global political strife and analyses the responses of Bali's business and community leaders to the crises that have buffeted the island since the fall of Suharto. In particular, it analyses crisis management with regard to the Bali Bombings, the impact of the bombings on the tourism development cycle and investigates the motives of the bombers. The authors argue that the actions of the bombers can best be understood with regard to the rise of political Islam as a global issue and breaks new ground with an analysis of the bombers' perspectives.

Tourism and Globalization

Surprisingly, in view of the size and scale of international tourism there are few publications that analyse this industry from the perspective of globalization. The lack is possibly due to the fact that tourism is a curiously fragmented industry – if it can be called an industry at all – since it involves many different kinds of businesses ranging from hotels and restaurants to airline and shipping agencies and a plethora of different interlocutors. Difficult though it may be to comprehend and interpret tourism's diversity, some serious attempts have been made to examine its global ramifications and perhaps some of the most successful of these have concerned islands, not least because tourism is so clearly important for many island economies, not least Bali. In particular, the Canary Islands seem to have attracted the most sustained analyses of the processes of globalization associated with tourism and, though they differ markedly in cultural, environmental and political terms from Bali, some common themes can be detected.

In his account of globalization in La Gomera, Donald Macleod reminds us that the island has long been part of a global trading network and thus part of a powerful geopolitical system since its colonization by Europeans. The author points out that the island continues this pattern by exporting its assets for tourism, but argues that the arrival of visitors who consume what they purchase in The Canaries inaugurates an entirely different set of relationships between the vendor and purchaser (Macleod 2004, 6). The fact that tourists consume at the point of production sparks off a whole series of different considerations of what is meant by globalization. Tourism may not be unique in this respect, but the fact that it does this on such a large scale deserves very serious consideration. One way in which Macleod finds helpful to think about this is to use the concept of the world-system theory (Wallerstein 1974; Shannon 1989) to interpret events both historically and as they continue today. The pivotal theme is that there is a core that controls the capital, and the levers of cultural and political dominance in relation to the periphery, which is exploited for its resources. The Canary Islands are thus peripheral to the urban core of continental Spain, and even within the islands themselves there remain a series of other smaller cores and peripheries (Macleod 2004, 7). These relations are by no means stable and what Macleod draws attention to is that the core has changed to include the metropolitan areas of Spain, Germany and the UK, with the European Union increasingly exerting its power in relation to development within the Canary Islands. Struggles between changing cores and peripheries are mediated through formal structures such as elected assemblies and politicians, and through informal but equally important ones such as patronage, set against a backdrop of increasing globalization.

What Macleod makes clear, however, is that the people of La Gomera are not simply passive recipients of an imposed transformation and indeed have become more actively involved with the process of globalization, notably tourism (Macleod 2004, 217). He reports moreover that the islanders have become engaged in a range of activities such as trading goods for export, migration, education, and in exhibiting elements of their culture for foreign visitors. Globalization in the Canary Islands seems to have led to the homogenization of economic and consumption patterns, while at the same time leaving room for a continuation of cultural elements such as the role of the family, language, religion, fiestas, music, arts, and crafts. These qualities help to create a shared sense of identity, which has been sharpened by the arrival of palpably different people. Macleod argues that on a global scale consciousness of such diversity leads to an awareness of '...greater wealth of difference – hence heterogeneity' (2002, 218).

It would appear that the cultural ramifications of globalization vary greatly since there are a number of other studies that place greater emphasis than Macleod on its homogenizing qualities. Ganjanapan, for example, points out that the process of globalization in Thailand since the late 1980s has transformed Thai society due to strong central control over the education and communication systems, leading to greater linguistic uniformity (2003, 132). The author goes further and argues that through the mass media even Thai identity itself has been transformed into a commodity and in the process, Thai-ness has become a signifier that is free from its specifically Thai ethnic and national essence (Ganjanapan 2003, 135). Ganjanapan reaches the conclusion that Thai-ness has achieved such universal status so that anybody no matter how remote from the Thai entity can indulge in Thai identity merely through the consumption of Thai-ness in the global market (Ganjanapan 2003, 136). This author does not specifically single out tourism, though presumably it is included in the general comments about the consumption of Thai identity.

In complete contrast Wayan Geriya argues that the impact of tourism on local culture in Bali is still very superficial, and that '...Balinese culture is strongly protected by local attitudes, social institutions, ways of life and sense of community' (Geriya 2003, 92). The interaction between tourism and culture has moreover is said to have led to cultural revitalization, but underlying these changes there are many continuities. But even such an apparently optimistic author does reveal a sense of unease when he reiterates the popular slogan 'Tourism for Bali, not Bali for Tourism' that suggests that any improvement and extension of tourism should maintain the existence and integrity of Bali (ibid.). It is, however, not only tourism that is an issue here since the policies advocated by the nation state within which Bali is located are clearly critical. Kagami argues, for example, that under Suharto's integrative regime (1966–1998) the Balinese made '...outstanding efforts to protect their local cultural forms and their identity' and won official support for their cause by arguing that their 'unique culture' was a major tourism resource that conveyed many economic benefits (Kagami 2003, 77–78).

Globalization in Bali is intimately linked to tourism, and a substantial literature has emerged decrying the threat posed by this industry to ancient harmonies, and 'untouched' and 'traditional' societies. Wood, however, laments the fact that tourism is often conceived of in terms of a game of billiards, in which the moving ball (tourism) acts upon an inert ball (the local culture). Wood's complaint is that this treats indigenous culture as uniform, passive and inert and he has argued that international tourism neither destroys local culture nor simply conserves it (Wood 1993, 66–7). Instead tourism is caught up in an on-going experience of cultural invention, in which Westernization is but a part of a wider process of cultural change. The new world system, instead of creating global cultural homogeneity, supplants one set of separate diversities with another set based on inter-relations. The salvation to these processes should not necessarily be sought in traditional ways of life because traditional approaches do not a priori guarantee workable solutions (Backhaus 1998, 170).

Tourism cannot readily be isolated from many other aspects of culture and the treatment of tourism as a solely exterior force may ignore how tourism can become an inseparable part of local reality (Picard 1993, 88–89). If culture is conceived of as static entity, then the actions and motivations of local participants are overlooked. Artistic styles, performing arts and even changes in dietary habits, can be seen as local attempts to accommodate the experience of tourism. Performances designed for tourists have, for example, been imported back into religious settings; Western theatrical conventions can be incorporated into sacred dances (Hitchcock and Norris 1995, 71). Performances created for national arts events and international audiences come to be regarded by the Balinese themselves as representatively Balinese. As Michel Picard points out 'touristic culture' has become so thoroughly internalised by the islanders that it contributes, paradoxically, to a reification of that which is conceived of as being authentically Balinese (Picard 1996, 199; Connor and Rubinstein 1999, 3). The culture that emerges reflects interaction with various participants, including the overlapping networks of tourism and the Indonesian state (Wood 1993, 67). As Yamashita has argued, it is not that Bali in its essence has survived without being tainted by Westernization and modernity and has come down to us intact, but that it has survived through '...flexible adaptation and response to stimuli from the outside world' (Yamashita 2003, 10).

Among the Balinese intelligentsia there is intense interest in the cultural influences associated with globalization and a series of books have appeared based on conferences in Europe, America, and Australia convened by international scholars on Bali and attended by Balinese scholars. These books have focussed the connections between globalization and modernity, tradition and the role of the media, especially television, and they include: *State and Society in Bali* (H. Geertz ed. 1991), *Being Modern in Bali: Image and Change* (Vickers ed. 1996), *Staying Local in the Global*

Village, Bali in the Late Twentieth Century (Rubinstein and Connors eds 1999), *To Change Bali, Essays in Honour of I Gusti Ngurah Bagus* (Vickers et al. 2000), *Inequality, Crisis and Social Change in Indonesia, The Muted Worlds of Bali* (Reuter ed. 2003) and *Bali and Beyond* (Yamashita 2003). Most of these studies focus on social change in Bali in the changing political and economic climate of reformist Indonesia, though Miguel Covarrubias had noted some of the changes taking place as early as the 1930s in his renowned *Island of Bali* (1937). Geertz's edited volume, for example, explores the changing relationship between state and society in Bali from various disciplines, ranging from textual studies of the genealogies of upper caste Balinese to the role of Balinese youth in the struggle to maintain Indonesian independence in the 1950s. Reuters's edited volume uncovers problems that would have been impossible to be studied in the Suharto era and includes papers disclosing the dynamism, as well as crises, in social relations within community, a legacy of the authoritarian regime. Vickers insightfully analyses how the Balinese have come to terms with modernity in the introduction to the volume he edited entitled *Being Modern in Bali: Image and Change* (1996). He discusses various forms of modernity in Balinese life and the excitement of the people in experiencing modernity during both the onset of colonialism and the awakening of nationalism (1996, 12–23). For the Balinese, Vickers argues, '...the issue is not that there is a single form of the modern, but that discourses of the modern are about who is authorised to act in Balinese society, that is, these discourses are part of a set of power relations' (1996, 35). The volume is distinctive because it analyses what modernity and especially globalization, especially what are perceived as global threats, mean in the everyday life of the Balinese. There is no doubt that the terrorist attacks in Bali represent the most immediate and tragic events associated with Bali's entrance into the global era. Through these attacks Bali found itself dragged into global imaginary clash between radical Muslims and the West. The 2002 Bali bombings sparked off radical social, political, and economic debates within Bali, not least on the true cost of tourism in terms of human suffering. Since those books on Bali mentioned above published before the Bali Bombings, the impacts of the bombings only appear in conference papers and journal articles. Besides dealing on history of tourism and its social impact, this book gives, however, considerable attention to the impact of the Bali bombings.

Public perceptions regarding tourism change in the wake of the bombings of 2002 and 2005, leading to criticism of the industry for dragging the Balinese into conflicts they have not initiated, and into compromising their cultural values through toleration of tourism's dark side, drugs, prostitution, money laundering and other shady businesses. Many Balinese interpreted the bombings as a sign that they had forsaken God and huge purification ceremonies were held to ask for God's protection. The blasts were said to be a warning, urging the people to return to their religion and traditions, which in this context explicitly meant agriculture. Farming is not only a source of sustenance, notably though the cultivation of rice, but is seen as a key cultural practice, 'agri-culture' being considered synonymous with 'Bali-culture' (MacRae 2005, 20). Perceptions on tourism are thus divided, whereas hitherto there had been more widespread agreement regarding its value for Bali, criticism is likely to endure and even intensify if tourism continues to flounder and not contribute sufficiently to Bali's prosperity. Members of the government and those who continue to earn a living through tourism are much more pessimistic about the potential of agriculture to deliver the harmony and prosperity that many Balinese crave. Highly skilled though Bali's farmers undoubtedly are, there are simply not enough jobs to go round in contemporary agriculture, and such has been the urbanisation of the island, notably in the south, that many Balinese, some several generations away from farm work, would find a return to the land hard if not impossible to adapt to.

Research Methods

This book grew out of a culmination of different but related research projects that were conducted between 1997 and 2006, and the methods employed were varied. The research on one of the highland villages was conducted during a series of short visits to Bali between 1997 and 2002, and reference was made to a questionnaire that was administered by 15 Balinese researchers drawn from a multi-disciplinary body known as the *Bali Human Ecology Studies Group* (Bali HESG). There were 35 questions in the questionnaire covering issues such as householder's occupation, the division of profits from tourism and how much the village's special status impinged on the lives of the local residents. The questions were written in a mixture of Indonesian (the national language) and Balinese that could readily be understood by the villagers, and the researchers, both male and female, asked the questions and then ticked the boxes of the questionnaire, sometimes adding hand-written notes on anything they found significant. The team completed a household survey (n = 76) of the village in August, 1999 and completed the research with a focus group meeting comprising members of the village assembly, and the results were written up in an unpublished article (Tourism Village of Penglipuran). The book also makes use of a survey and a focus group meeting held in Benoa village in 2004, that were undertaken as part of a project entitled 'Building Research Methods in Pro-Poor Tourism', which was funded by the ASEAN-EU University Network Programme.

The methods used in collecting data on street traders and hotel ownership patterns were essentially ethnographic, involving a hybrid approach that may be characterised as 'fast and dirty'. Known as Participatory Rural Appraisal this approach is derived from social anthropology and has been adapted for use in development studies, and more recently tourism studies (Hampton 1998, 645). Following Michaud (1995, 682), this approach might be termed 'pre-fieldwork' since it involves visits lasting several weeks as opposed to participant observation conducted over a much longer period. It is suggested that this approach is applicable to the study of entrepreneurs in tourism since, although the contexts are not strictly rural, the settings pose similar problems. The work on the informal sector the work was conducted through the spontaneous questioning of street traders and shop and restaurant employees, as well as expatriates living in Sanur, one of Bali's main resort areas. A survey of this kind was also conducted in one of Bali's offshore islands, Nusa Penida, in 2004.

A variety of other qualitative methods were used to gather information, including spontaneous questioning in the national language, Bahasa Indonesia, semi-structured informal interviews and more detailed repeat interviews with key informants. In particular the research on the Bali bombings was undertaken in two main blocks of about a month's duration in July 2003 and July 2004, though work continued intermittently in between the main phases up until 2006. In view of the sensitivity of the issues raised, not least the background of religiously inspired terrorism, it was not possible to administer a questionnaire, in the manner of previous studies (for example Cukier and Wall 1994), and all informants remain anonymous. The research on the formal sector comprised semi-structured interviews (n = 30), usually in the informant's office, with Balinese tourism officials, the leaders of Indonesian trade associations and general managers of hotels. The language barriers were minimal since Balinese is spoken by the Balinese co-author, and both researchers speak and read English and Bahasa Indonesian, the widely spoken national language. The authors also attended workshops in September 2003 designed to help the recovery of Bali's tourism industry, which were attended by representatives of local government, industry, NGOs, academia and the media, the main local stakeholders in Bali's tourism industry. The authors of this study also made use of Indonesian newspaper coverage of the crisis and had access to the trial transcripts of the alleged Bali bombers since some of their responses during cross examination also have a bearing on inter-ethnic relations.

The authors also examined the creative literature in Bali, particularly with regard to Balinese-foreign relations, and thus there is a significant element of literary criticism within the book. Reference is also made to a variety of media, including popular commentaries, and in particular we made use of the *Bali Post*, the locally published newspaper that is read by comparatively well-educated islanders. The newspaper cannot be said to be genuinely representative of popular discourse on the island, but it does serve as a valuable record of the dates of events and the activities of officialdom, and provides insights into the perspectives of the island's opinion leaders. Picard has characterized these people as personnel of the provincial government and the intelligentsia at large – academics, journalists, bureaucrats, technocrats, entrepreneurs and professionals – and though they do not necessarily share the same opinions, they do tend to live in and around the capital, Denpasar. These opinion leaders mediate between the village and the state by speaking on behalf of the Balinese to Jakarta, and by conveying the national capital's ruling to the province; by doing so they simultaneously affirm their Balinese identity and Bali's integration within the Indonesian state (Picard 1997, 207).

Focus of the Book

The book aims to provide an in-depth analysis of tourism as a global phenomenon in the 21st century in relation to one of the world's top tourism destinations looking at its impact on ethnicity and identity. It provides one of the few accounts of the impact of the Asian Crisis on tourism and the longest empirically based analysis of the Bali Bombings, including new material that is not yet in the public domain. This study provides a detailed analysis of the only popular revolt against the introduction of a World Heritage Site in Asia and uniquely discusses UNESCO's project from the perspective of globalization. The book charts the rise of tourism through its introduction by colonial agency and its huge rise under an authoritarian regime, discussing the role played by multi-national hotel chains and the emergence of local entrepreneurs who are on the threshold of operating globally. The book shows how indigenous traditional institutions can cope successfully with a series of globally-induced crises and offers a robust theoretical analysis of the role of tradition in management. The approach is multi-disciplinary and combines a social science analysis of social change and development with a humanities-based approach to textual investigation of both creative writing and the contemporary media.

Chapter 2

The Importance of Tourism

The Island of Bali

Despite its small size (5,632.86 square kilometres) relative to major Indonesian islands such as Java and Sumatra, Bali is densely populated with over 3 million inhabitants. The island is divided by a mountain range running east to west, the highest point being the volcanic peak (3,142 metres) of Gunung Agung. The majority of the population lives in the south-central region, on the slopes that run from the mountains to the southern coastline. The Balinese tap the numerous rivers running from the uplands to the sea to create verdant rice terraces, which can yield up to three harvests per year, long the mainstay of the island's economy and the basis of its civilization. Bali also marks the point where the Asian ecological zone starts to give way to the Australian one and is divided from neighbouring Lombok by the Wallace Line, named after Sir Alfred Russell Wallace, the co-originator with Darwin of the theory of evolution through natural selection.

The Balinese language belongs to the Austronesian family, which embraces the majority of the languages of maritime Southeast Asia, as well as the languages of the indigenous Taiwanese, the peoples of Oceania, with the exception of the Papuans and Australian Aborigines, and the earliest settlers of Madagascar. Diverse those these people may be, they have much in common in cultural terms, notably a strong belief in the spiritual power of the ancestors. Bali, however, was strongly influenced by two of Asia's world religions, Hinduism and Buddhism, through contact with neighbouring Java and by direct trade relations with India. In accordance with the state code known as Pancasila all Indonesians are expected to subscribe to a recognized world religion, and thus the Balinese describe themselves as Hindu, though it is perhaps more accurate to describe them as Bali-Hindu, since many local elements have been incorporated into their belief system and this sets them apart from their co-religionists in India.

By the late tenth century the ruling households of East Java and Bali had become closely interconnected as a result of the marriage of a Balinese prince, Udayana, and a Javanese princess Sri Gunapriya Dharmapatni (Hobart, Ramseyer and Leeman 1996, 27). In 1284 Bali was subjected to Singhasari rule from East Java, after which there was a period of renewed independence that ended with incorporation into the powerful Hindu-Javanese kingdom of Majapahit (ibid., 33). Towards the end of Majapahit rule (*c.* 1500) Islam became the dominant faith in Java, but Bali remained steadfastly Hindu. According to legend, refugees from Java who refused to convert to Islam fled to Bali, where many leading families to this day still describe themselves as *Wong Majapahit* (people of Majapahit) (ibid., 38).

Shortly after the triumph of Islam, Java began to be subjected to European colonial incursions that led eventually to the establishment of the Dutch East Indies, but Bali remained largely independent until the early 20th century. The colonization of the island was a protracted affair, starting in the north of the island in 1846, and culminating in the suicidal battles of 1906 and 1908 in the south. The Dutch triumph was, however, overshadowed by critical press coverage of the massacres of the royal households who resisted them; accompanied by their families and retainers the rulers of Badung and Klungkung marched on to the Dutch guns in a fight to the death, *puputan* (ibid., 202). Reporters and photographers accompanying the Dutch expedition were at hand to witness these appalling events and their accounts severely Holland's image as a responsible and humane colonial power. Dutch hegemony collapsed with the Japanese invasion of 1942, but their rule was short-lived and by 1945 they had begun surrendering to the Allies. Fearing a return to colonial rule, Sukarno, whose mother was Balinese, proclaimed Indonesia's independence on 17 August, 1945 along with his fellow nationalists, Mohammad Hatta and Mohamad Yamin, but the Dutch were not effectively ousted until 1949 when they agreed to recognize the new republic at the Round Table Conference in the Hague (Hanna 1976, 119). It is only recently that the Dutch government has begun to acknowledge 1945 as the year of Indonesian independence.

Since independence the Balinese have been citizens of the Republic of Indonesia, though they have retained a very strong sense of their own identity as almost a state within a state. The Balinese themselves are divided internally between those who follow the old religion, Bali Hindu, almost 92% of the population and those who profess Islam, Bali Islam, the smallest minorities of all being those who subscribe to either Buddhism or Christianity. The situation is reversed in the rest of Indonesia where Muslims comprise the overwhelming majority, and thus Bali is often described as a Hindu island in a Muslim sea. This, however, is a little misleading since there remain Hindu pockets in East Java, as well as a sizeable Hindu minority in neighbouring Lombok. As a result of Indonesia's controversial transmigration programme, whereby settlers from the more densely populated islands are shipped to more sparsely settled ones, there are now sizeable communities of Balinese in Lampung, South Sumatra, Sulawesi and elsewhere. There is also a small Chinese minority in Bali whose numbers were swelled as a result of the anti-Chinese riots of May 1998, but it remains unclear how many eventually took up permanent residence on the island.

The idea that Bali is somehow qualitatively different from the rest of Indonesia is reinforced by the tourism promotional literature that often ignores its status as an Indonesian province. The island is often described as the 'land of a thousand temples', which is a rare understatement in the hyperbolic world of tourism marketing, since the Department of Religious Affairs acknowledges 4,661 temples, excluding minor ones. This sense of separateness is also apparent in airline terminals where 'Bali' is often referred to as the destination as opposed to Denpasar, the island's capital city; flights to other Indonesian cities are invariably advertised under their own names. Despite these perceptions Bali has been an integral part of Indonesia since independence, albeit the only Indonesian province whose territory embraces a complete island (Picard 1993, 92). Anomalous though Bali is, it is still not accorded special status by the Indonesian authorities like the provinces of Jakarta, Aceh (Sumatra) and Yogyakarta (Java).

Colonialism and the Introduction of Tourism

Tourism development in Bali became a priority for the government of the Netherlands East Indies partly to atone for the massacres that accompanied the imposition of colonial rule. Severely criticized back home in The Netherlands for the killing of civilians, the colonial regime introduced more ethical policies such as the building of schools and the introduction of measures to help preserve local culture and develop tourism (Picard 1997, 185). This may have been an economically reasonable strategy since the island lacked an abundance of lands suitable for colonial plantations and produced few export commodities (Boon 1977), but it was also Dutch policy to prevent big companies from opening up rubber or tea plantains or sugar or tobacco estates as had happened in Java. Only a few Dutch businesses therefore found it profitable to trade in Bali and they were mainly concentrated around the urban areas of Buleleng and Denpasar (Hanna 1976, 103). There was also an absence of any conspicuous colony of Western residents – evidence perhaps of a more ethical outlook – and, though Westerners did settle in the island, notably the artists, scholars and writers of the inter war years, they were not concentrated in one location.

The origins of tourism can be traced back to 1908, a few weeks before the last Balinese kingdom fell, when representatives of commercial banks, railroads, insurance companies and shipping agencies founded in Batavia (now Jakarta) the Association of Tourist Traffic in Netherlands India. The founders included the Royal Packet Navigation Company (KPM), which had a shipping monopoly in the Dutch East Indies (Picard 1996, 23). This government-subsidized conglomerate, which had established relations with the principal tour operators of the period, opened an Official Tourist Bureau the very same year. Their activities were initially limited to Java, but had extended their operation to Bali by 1914, marketing the island in brochures as the 'Gem of the Lesser Sunda Isles' (Picturesque Dutch East Indies 1925).

The Dutch also implemented a policy known as Balinization (*Baliseering*) in the 1920s that was designed to make Balinese youth conscious of their rich heritage through an approach to education that placed emphasis on the study of their language, literature and traditional arts, while discouraging inappropriate expressions of modernism (te Flierhaar 1941). The colonial authorities organized collections and inventories of Balinese artifacts, not only to make them accessible to scholars, but also to stop them from being sold as souvenirs to foreign visitors. With hindsight one might be tempted to regard these policies as enlightened by the standards of the period, but in reality what was being attempted was an early form of what came to be known as 'social engineering'. The key point is that the Dutch were not so much interested in preserving the culture of Bali as they found it, but in restoring it to what they thought was its original integrity. It was also an extraordinarily contradictory stance since they were on one hand try to shield the islanders from outside contacts while simultaneously developing tourism.

In their attempt to attract tourists and to show off the successes of their benevolent rule, the Dutch spared no effort to expose what they saw as the glories of Bali's civilization to international audiences. Publications of the period from the Tourist Bureau show how Bali's image of a tourism destination was evolving. The first publications were characterized by their restrained and essentially practical nature, usually with little information about the history and culture of the island. But by 1927 the Tourist Bureau had begun publishing more detailed accounts, including notices about Balinese festivals, as well as cremations, for which the Bureau was prepared to charter a ship if it were to be a particularly spectacular one (Picard 1996, 25). One of the most important showcases was, however, the Colonial Exhibition in Paris in 1931 where dancers and musicians from Ubud, and the neighbouring village of Peliatan performed under the leadership of Cokorda Gede Raka Sukawati, the lord of Ubud. The staging of Balinese arts in such a global forum away from the island was an historic moment for the Balinese. To underscore their serious intentions, the Dutch entrusted one of their foremost experts, Roelof Goris, with the editing of a small booklet entitled *Observations on the Customs and Life of the Balinese*, which was re-published in a revised edition in 1939. As tourism became established, the number of publications grew and publications became more lyrical and gushing about the beauty of the island and its inhabitants.

The first tourists to arrive travelled around the island by horse or by car, staying overnight – space permitting – in the government guesthouses used by colonial officials on their tours of inspection around the island (Picard 1996, 24). Following the introduction of a regular steamship service in 1924, linking Buleleng (the port serving Singaraja) to Surabaya, Semarang, Batavia and Singapore, the number of tourists began to rise, sparking off a mini tourist boom that stagnated a little in the depression years of the early 1930s (Picard 1996, 25–7). In 1928 the KPM opened the Bali Hotel, replacing the Denpasar guesthouse, in the vicinity of the *Puputan* of 1906, shortly after renovating the guesthouse in Kintamani, which was thereafter reserved exclusively for visitors who wished to enjoy the spectacular panorama over the crater lake of Batur (ibid., 24). It was not long before cruise ships began docking at the island, though the colonial authorities had to upgrade the port facilities in the Padangbai in the southeast of the island in order to receive them. By the end of the 1920s an average of four ships a week was serving the island, and by 1934 a daily ferry service was in operation between Gilimanuk in the west of the island and Banyuwangi in eastern Java. Air services were also introduced, linking Surabaya to Denpasar in 1934, leading to a purpose-built airport at Tuban in 1938, which handled three flights a week.

It is difficult to determine precisely how many tourists came to Bali in the early days of tourism because they are not specifically identified as such in the lists of registered visitors. The first figures published by the Tourist Bureau recorded 213 visitors in 1924, rising rapidly to 1,428 in 1929, but declining in the early 1930s. Numbers recovered in 1934 and went on to reach an average of 3,000 per annum by the end of the decade (Picard 1996, 25). Bali's hotel capacity in the inter war years comprised 64 double rooms divided between 48 at the Bali Hotel and 16 at the Satrya Hotel, a Chinese owned concern that was built in Denpasar at the start of the 1930s. By 1936 tourists who had become bored with the urban comfort of Denpasar could sample an entirely different type of accommodation, bungalows built in the Balinese style, close to the magnificent beach of Kuta. The American artists, Louise Garrett and Robert Koke, built the first of these under the name of the Kuta Beach Hotel, though a second hotel, the Suara Segara (the Sound of the Sea) appeared shortly afterwards. The latter was developed by the notorious K'tut Tantri, who later found fame as a propagandist for Sukarno under the pseudonym of 'Surabaya Sue', and who initially worked for the Americans only to become their fierce rival.

Little is known of the role played by the local community in tourism development on the island during the colonial period, though there Vickers reports on the involvement of people in North Bali as tour guides and souvenir sellers (Vickers 1989, 115). An exception is the foundation of the art society known as Pita Maha in Ubud in 1936, which is often not associated with the development of tourism, though tourists were one of the main markets for its products. Convened by Bonnet, Cokorda Gde Agung Sukawati, Walter Spies, Cokorda Gde Rai and I Gusti Nyoman Lempad, the society started to sell paintings and sculpture to tourists, though the finest pieces were reserved for the museum collection (Hilbery 1983, 21). Regular performances for tourists called the Bali Night were also arranged Cokorda Gde Agung Sukawati and Walter Spies in Ubud, probably incorporating some of the experiences gained by dancers at the Colonial Exhibition in Paris. In order to attract visitors to Ubud, Cokorda Gde Agung Sukawati used to travel regularly to Denpasar to lure tourists into staying in the Bali Hotel; he also used these experiences to improve his English.

The most important thing from a Balinese perspective was the move to support colonial attempts to maintain Balinese arts and culture, and to make them as attractive as possible for tourists. In particular, the participation of a Balinese arts troupe in the Paris Colonial Exhibition can be seen as a local contribution to tourism promotion. At home, the islanders organized themselves into groups of artists in order to put on regular performances for visitors at venues such as the Bali Hotel. From this perspective, one may argue that Balinese were willing participants in tourism development, though one should bear in mind the colonial context. People from other indigenous ethnic groups contributed to the promotion of the island, notably through the publication of a guidebook in Malay for a national audience. In this respect, it is worth mentioning the contribution made by a Chinese-Indonesian, Soe Lie Piet, who published at least two guide books on Bali.

It is worth noting, however, that the Balinese also had a critical view of this new industry and there is some evidence from the 1920s and 1930s, which suggests that the islanders began to express their reservations about the negative impacts of tourism. Concerns were expressed in local publications such as *Surya Kanta* and *Djatajoe* and attention was drawn, for example, to roads damaged by frequently passing vehicles carrying tourists as damaged road frequently and a road left un-repaired because it did no connect with a tourism destination. Another issue that was highly criticized by Balinese authors was the habit of tourists taking photographs of bare-breasted women and they urged the colonial government to ban such practices (Darma Putra 2003, 33–6).

Regardless of what certain critics felt, Bali continued to be marketed worldwide by various agencies, but it was not their activities alone that helped establish the island's international reputation. Western scholars, painters, sculptors, film makers and photographers who sojourned on the island in the inter war years helped to raise Bali's profile, not least through their publications that included travel books, fiction and more serious works, many of which remain commercially viable in the 21st century. Those that can still be picked up in bookshops in the resorts of South Bali include: Gregor Krause's photographic study, *Bali 1912*, Vicky Baum's *Tale from Bali*, and Covarrubias' *Island of Bali*. The visit of Pandit Nehru, the first prime minister of India, helped to boost the island's prestige; not content with just praising its cultural and natural beauties, he described Bali with a slick hyperbole as the 'the morning of the world'.

From the outset Bali was not simply an international tourism destination since there was an important domestic market to consider comprising colonial officials and business people, and the members of the indigenous elites. There seems to have been a ready market for publications about the island in Malay, the forerunner of the national language of Indonesian, including poems, guidebooks and novels. The Chinese Indonesian writer, Soe Lie Piet, for example, set two novels in Bali that drew on anthropological themes to romanticize the island.

Of more direct importance are Soe Lie Piet's guidebooks entitled *Pengantar ke Bali* and *Pengoendjoekan Poelo Bali Atawa Gids Bali*. These books do not contain details of year of publication, but it seems likely that they were written in the 1930s. The books provided descriptions of Balinese culture, tourist attractions, and were illustrated by photographs of panoramic landscapes, temples, dances, other aspects of cultural heritage and the ubiquitous bare-breasted women that presumably appealed as much to Asian visitors as they did to Westerners. In the introduction the author reported that many American and European tourists had visited Bali, almost as if this was a measure of how important the island was. He then goes on to say that while many books had been published about Bali in foreign languages for international visitors, so far there had been none in Malay. He wrote:

> Meliat ini kakoerangan, sedeng djoemlahnja orang-orang jang djalan-djalan ka Bali dari saloeroeh Java, Sumatra, Borneo, Celebes, enz. jang berbahasa Melajoe tjokoep banjak, maka saja merasa ada mempoenjai itoe kewadjiban aken bikin satoe boekoe goena maksoed terseboet – jang barangkali sadja moedah-moedahan aken mendjadi satoe kagoenaan bagi sekalian pembatja dari Indonesia.
>
> Seeing this lack, while the number of people travelling to Bali from all over Java, Sumatra, Borneo, Celebes etc., many of whom speak Malay, was sufficiently great, I felt an obligation to write a book to fill the gap – with the hope that it will be useful for all Indonesian readers (1936: 5).

What Soe Lie Piet and other local writers did was to introduce Bali as a tourism destination to the indigenous peoples of the Netherlands East Indies, much in the way that foreign writers opened up the island to the international community. Despite the involvement of locals, tourism was at this stage a largely foreign induced initiative.

The Post War Era and Sukarno

During the colonial and interwar periods external agents played a more significant role in Bali's tourism development than local concerns, and even after independence expatriates continued to play a significant role. There were, however, signs that the Balinese themselves were starting to become more actively involved, one of the first local hoteliers being the King of Ubud, Cokorda Gde Agung Sukawati, who opened up his palace for paying guests. But, as the King himself admitted, the idea came first from the painter Rudolf Bonnet (Hilbery 1983, 53–54). From then on, Cokorda Gde Agung Sukawati, expanded his provision of accommodation services. Ubud began to receive visitors from around the world, though the numbers remained small. Other significant locals who became involved in providing hospitality in the 1950s included Ida Bagus Kompyang (1927–) who established the Segara Village Beach Hotel in Sanur in 1956, ten years before the construction of the more famous Bali Beach Hotel. In Lovina, the late literary figure, Anak Agung Panji Tisna (1908–1978) built accommodation in Lovina, one of the most popular beach destinations in North Bali.

With regard to travel and transportation, the contribution of a former local teacher I Nyoman Oka (1911–1993), better known locally as Nang Lecir, is worth noting. He and other local businessmen, as well as political activists like I Gusti Putu Merta, set up the first locally owned travel company in 30 April 1956 called NV Bali Tours. This company requested permission to open its office at the Bali Hotel, but was rejected due to lack of space. Despite this setback, the company was permitted to conduct its business activities around the hotel by operating taxi and bus services, and selling tours to hotel guests. By the end of the 1950s, the company was well established and was actively cooperating with 13 travel agents from overseas. But, the fate of NV Bali Tours was sealed when one of Sukarno's ministers issued an edict in a letter dated 30 March 1960 that only permitted the Jakarta-based travel agent, Nitour, with a branch in Bali to handle international tourists. Nitour was given a monopoly of the foreign and more lucrative trade, while the locally owned travel agency was encouraged to restrict itself to domestic tourism, a somewhat limited market at that time. Central government had thus succeeded in snuffing out local competition, a foretaste of what was to come under Suharto, and, though NV Bali Tours was given the chance to unite with Nitour, the offer was rejected by Nang Lecir in the name of Balinese self-esteem (Darma Putra 2003a, 4–5). The name of Bali Tours eventually reemerged, though owned and operated by different people, but Nang Lecir had the consolation, along with the three aforementioned local hoteliers of being presented with awards, *Karyakarana Pariwisata*, by the provincial government of Bali for their pioneering achievements.

Despite the turbulence of the Sukarno years, the newly independent Indonesian government continued to promote Bali abroad as a tourism destination, which was ironic given the President's relentless anti Western rhetoric. This apparent contradiction can be explained by the fact that Sukarno retained a deep affection for the island, not least because his mother was Balinese. Bali's culture came to stand for Indonesian culture in promotional terms, just as it had under the Dutch, and Balinese dancers were often invited to perform in the presidential palace in Jakarta and to accompany foreign missions to the Soviet Union, Iran, and Singapore. In particular, John Coast, who was close to Sukarno, took a Balinese dance troop from Peliatan (Ubud) on a successful tour of the United States in 1952–53. The show seems to have captivated American audiences and it made the front cover of *Dance Magazine* in September, 1952, and was featured in editorials in the *New York Times* and *Herald Tribune* as heralding a new era in East-West cultural relations (Coast 1966).

The fact that tourists continued to visit Bali, despite the growing leftwing activism in Indonesia, is due to Sukarno's ongoing endorsement and promotion of his mother's homeland. It is sometimes assumed that visitors were put off by the Indonesian government's stance, but comments from the period in the guest book of the palace in Ubud, Puri Saren, paint a more pleasant picture. The custom of signing the book was instigated by Cokorda Gde Agung Sukawati, and he succeeded in soliciting a great deal of feedback, which was invariably enthusiastic. The visitors included artists, lecturers, bureaucrats, diplomats, heads of state and journalists, and their comments, sometimes taking the form of poems and sketches, extolled the beauty of Bali and the hospitality and creativity of its people.

In view of the fact that Sukarno took such a personal interest in the island and acknowledged his Balinese heritage meant that support for developing tourism became more widespread on the island. The Balinese also began to become involved in tourism in a variety of ways such as tour guides, transport providers and souvenir sellers. Sukarno relished taking state guests to the Tampaksiring State Palace, and Ubud, and as a result the island continued to receive a lot of free publicity. Using war reparations from Japan, Sukarno built the Bali Beach Hotel in 1966 and started the expansion of Ngurah Rai Airport, which eventually opened in 1969 when Suharto had replaced him as president. The Bali Beach Hotel, the only ten-storied hotel on the island, and the Ngurah Rai International Airport were not only the two most important tourism installations on the island, but they came to represent Bali's increasing involvement in international tourism, as explicit icons of globalization

Suharto's Tourism Boom

Despite the antipathy between Indonesia's first two presidents, Suharto continued to build on Sukarno's developments in Bali. Realizing that a major hotel without an international airport would not be viable, Suharto carried on providing funds from central government for the benefit of Bali. Sukarno may have given the go ahead for Bali's two biggest tourism projects to date, but he was not to see them through to fruition because his political power was waning. He was not invited to preside over the official opening of the Bali Beach Hotel and in his place Sultan Hamengku Buwono, Suharto's Vice President, did the honours. The pace of globalization began to pick up once the international airport opened and very soon Bali saw its first hotel managed by an international chain, the Intercontinental. After the haphazard style of planning under Sukarno, attempts were made to adopt a more systematic form of development, and tourism was included in the first Five Year Plan (1969–1974). Shortly afterwards in 1971, a Paris-based consultancy SCETO put forward its grand design for tourism development on the island, especially the dry and infertile coastal area of Nusa Dua.

Development spread, and in 1983 Garuda, the Indonesian national airline, opened the first property in Nusa Dua called Nusa Dua Beach Hotel. This resort on the southern tip of Bali soon proved to be a magnet for a variety of international hotel chains and soon Club Med, Hilton, Hyatt, Sol, and Sheraton had become established. The Suharto government's decision in 1983 to introduce free visas for tourists helped to fill up the rapidly expanding accommodation sector in Nusa Dua and Kuta, and the hosting of Ronald Reagan at the Nusa Dua Beach Hotel in 1986 provided much needed publicity and served as a draw for Australian and Japanese visitors. A decade later, in 1991, in order to get more fresh cash to expand its company, Garuda Indonesia sold the Nusa Dua Beach Hotel to the Sultan of Brunei. The rapid expansion in tourism arrivals led to a renewal of development in Kuta as the former small *losmen*, and pension-type accommodation was replaced with star rated hotels like Shangrila, Ramada and Hard Rock Hotel.

Apart from visiting dignitaries, a variety of other events began to be used to promote tourism, notably the Bali Ten 10km, which was an international sports event attended by champion runners from Africa, America, Asia, and Australia, cheered on by thousand of tourists on holiday in Bali. The event was initiated in 1986 by Bob Hassan, an Indonesian-Chinese businessman close to Suharto and his minister during the last years of Suharto's rule, succeeded in attracting widespread international media coverage to the benefit of tourism. The running contest moved to Borobudur, while in Bali it continued under the name Bali Nittoh Marathon. When held in Nusa Dua, the event, which was sponsored by a Japanese tea company, Nittoh, proved to be very popular among Japanese tourists and helped to boost the number of Japanese visitors to the island in the mid 1990s.

From the late 1980s to the mid 1990s tourism boomed in Bali, and investment flowed into the island to build hotels and restaurants, and to open a cruise ship company. The number of international airlines flying to Bali rose sharply to include, in addition to the national carrier Garuda Indonesia, KLM, Lufthansa, UTA/ Air France, Lauda Air, JAL and ANA from Japan, Singapore Airlines, Cathay Pacific, Malaysia Air Service (MAS), Brunei Air, Air New Zealand and Thai Airways, as well as the Australian airlines Qantas and Ansett Australia and the American based-airline, Continental Micronesia. Not all of these operators survived the onset of competition from budget airlines, but they all contributed to widening global access to Bali.

The infrastructure of tourism spread even further afield to include Sawangan (south of Nusa Dua), Tanjung Benoa (north of Nusa Dua), Jimbaran, Pecatu, and Canggu, though one development came to symbolize growing local disenchantment with tourism's relentless expansion, the controversial hotel and golf course at the temple of Tanah Lot (southwest Bali). The Bali Nirwana Resort and golf club managed by Le Meridien faced strong local resistance to what was seen as insensitive development so close to an especially attractive temple, though the will of Suharto's government eventually prevailed, and the resort went ahead. The involvement of yet more hotel chains helped to publicize Bali globally and at the same time the Jakarta-based conglomerates, some with close links to the President and his family, started to open up hotels. Commentators in Bali started to describe the island as a colony of Jakarta, but at the same time local Balinese entrepreneurs began to play an increasingly important role to bring the name of Bali into the global market, and this helped ameliorate some of the criticism.

By 1997 the onset of the Asian Crisis had started to undermine Suharto's authority and the following year he was forced to step down in the face of mass demonstrations led by students. As Indonesia plunged into a series of crises that quickly became known as the 'total crisis' (see Chapter 8), tourism arrivals began to drop nationally, but Bali continued to enjoy a relative calm, and the tourists, especially the Australians, kept coming to enjoy the benefit of the strength of the Australian dollar against the Indonesian rupiah. Their ongoing presence underscored how important global tourism had become for the island's economy.

From Total Crisis to Terror

Habibie replaced Suharto as president and during his interregnum he allowed the people of East Timor to vote on whether or not they should remain part of Indonesia. When the inhabitants of this former Portuguese colony, which had been annexed by Suharto in 1975, voted overwhelmingly in favour of separation from Indonesia, the military reacted badly. Militias loyal to Indonesia and elements within the army ransacked East Timor, destroying much of its infrastructure. Despite Bali's proximity to eastern Indonesia and persistent rumours of militiamen passing through the island on their way back from East Timor, tourism remained reasonably buoyant. It was not until the end of Habibie's presidency that Bali experienced serious public disorder. After the parliamentary elections, in which Megawati Soekarnoputri's party, the Indonesian Democratic Party, won the largest share of seats, it was widely expected that she would become president. At that time, however, presidents were elected by parliament and not by a direct vote, and Abdurrahman Wahid – or Gus Dur as he is popularly known – emerged as the winner. The news was greeted with riots in Bali where there is strong support for Megawati's party, though the perpetrators were not known; the unrest shocked the Balinese and alarmed the tourism industry, but the impact on visitor arrivals was limited because the problem was quickly resolved, and the international media were quick to spread the news that Bali was back to normal.

Like her father before her Megawati continued to promote Bali for tourism, but the bombings of 2002 undermined these efforts, leading to the sharpest decline in international visitor arrivals since records began. The industry was further unnerved when the immigration authorities decided to introduce visas on arrival. There were worries that the introduction of more controls would inhibit the recovery of visitor arrivals, and there was widespread opposition to the new visas. Their fears appeared to be unfounded since by 2004 Bali's tourism had started to recover, but whether the upswing would have been stronger without the new controls remains a moot point. The immigration authorities were adamant that the new visas were necessary for reasons of security and revenue raising and the controls remained in place after the succession of Susilo Bambang Yudhoyono. A second round of terror attacks in Jimbaran and Kuta on 1 October, 2005, led to another dramatic drop in tourism arrivals, leaving the tourism sector in big trouble once again.

From the outset it is clear that globalization and tourism have been closely interlinked in Bali. As the air links made Bali more accessible more hotel chains were attracted and there was a knock-on effect on other industries, notably in the arts and crafts sector, and the garment industry. From the business point of view, Bali has surely turned itself into a widely recognized brand and this appears to have accelerated the process of globalization as more multinationals have sought to profit by association. At first glance it would appear that Bali became caught up in a virtuous circle with more tourists wanting to visit, as the brand became better known, and more multinational companies wanting a share of the action as the economic situation improved. Many Balinese have welcomed these changes, but the downside is all too apparent. On one hand there have been concerns about tourism's impact on Balinese culture, one of the main draws for tourists, and on the other there are worries about the environmental degradation associated with this industry on what is after all a relatively small island. To top it all, Bali's iconic status has attracted the attention of terrorists looking for somewhere to make their cause more widely known while at the same time striking a blow against Western interests by attacking tourists. If the island had not been a tourism destination of worldwide repute, then it seems unlikely that the terrorists would have shown much interest in it, though its status as a Hindu enclave might have made it vulnerable at some stage. The fact that Bali was close to East Java, the home of bombers such as Amrozy also seems to have been a factor. Significantly, globalization seems to have replaced an earlier awareness of tourism in Balinese discourse as a source of innovation and change: '...exciting but potentially threatening' (Rubinstein and Connor 1999, 3), and the terror attacks of the 21st century have heightened these sentiments.

Importance of Tourism

Studies by Indonesian scholars underline the importance of tourism in the Balinese economy, though we should not lose sight of the ongoing importance of agriculture (Dewa Made Tantera and Agus Pakpahan 1990, 189–191). As can be seen through the provision of 'homestays', tourism can provide farmers with an additional source of income, helping them to diversify. Inexpensive guest rooms can be added to rural dwellings and, because of the low overheads, have the potential to remain cost effective even during lulls in the tourist cycle. In material terms, Bali has been in the late 20th and early 21st centuries one of the wealthiest Indonesian provinces, and this is due as much to its tourist industry as to its highly skilled farmers. The introduction of infrastructure associated with tourism, such as roads and airports, has facilitated the growth of small factories, turning Bali into a major handicraft-exporting centre. The demand for high quality Balinese-made goods, especially in interior design and fashion, has undoubtedly helped many Balinese businesses to flourish. The skill of Balinese craftsmen may be seen not only as a reflection of the vitality of Balinese culture, but as a rich source of added value.

Foreign leakages whereby the money gained through tourism flows out of the country to purchase the imported goods and services used by tourists do occur, and to these outflows of capital may be added the transfers made by foreign and jointly owned businesses seeking to repatriate profits abroad. But as Hugh Mabbett (1987) has pointed out, the situation is by no means as bleak in Bali as it appears to be, though his unqualified optimism overstates the case. The sprawling resorts of Kuta and Sanur still have a high proportion of locally owned businesses and have given rise to a generation of local entrepreneurs, many of them women. Developments in Kuta and Sanur may appear unsightly, but there have been successes, such as the application of a tree-line policy, which restricts the height of hotels to that of the tallest coconut palms. In comparison with other Southeast Asian resorts, such as Pattaya, Kuta and Sanur are relatively restrained.

As is the case elsewhere in Southeast Asia, there are environmental problems associated with tourism that need urgent attention. Tourists consume large numbers of shrimps, for example, and the farms that help to satisfy this demand may alter the ecology of agricultural land, making it difficult to return to rice farming when the investors decide to move on elsewhere (Backhaus 1998, 190). Bali also has reefs that were damaged during the hotel building booms of the 1980s and early 1990s when coral was removed to make lime for the construction industry. As tourists begin to look for new attractions inland away from the crowded coasts, effective visitor management will be needed for destinations that lie off the beaten track, such as the wildlife reserve of Negara in the west of the island.

Increasingly, the debate concerning tourism in Bali has turned to questions of sustainability, a particularly pressing question given the downturn in arrivals occasioned by the bombings. There is also competition from other Indonesian islands such as Lombok, which has long been promoted as being like Bali used to be in the 1960s. Then there are international competitors that try to steal Bali's clothes, such as the Maldives advertising the use of Bali-style bungalows in its resorts. When a tourist destination like Bali becomes a kind of archetype of the tropical island holiday experience, then there is likely to be some image homogenization in the eyes of would be holiday makers trying to decide where to go and be offered yet more beaches, waving palms and colourful festivals. There is the debate concerning the quality and economic status of the tourists with complaints from general managers of hotels that the Asian budget travellers, mainly from Taiwan and Korea, who have taken advantage of falling prices in the aftermath of the bombings are not especially profitable since they skimp on using the services provided by hotels, notably restaurants. A low volume, but high spending tourism industry might be more appropriate to the island's needs than mass tourism, although the needs of low budget tourists using 'homestays' also need to be taken into account. A further consideration is that as a result of the political reforms since the fall of Suharto, the Balinese are much more inclined to protest against the tourism developments that they object to, but these protests need to be analysed on a case-by-case basis and should not be taken as a generalized anti-tourism stance on the part of the islanders. Tourism continues to be warmly welcomed, but increasingly discussions about its sustainability are set within the context of globalization.

Chapter 3

A Brand Created

Paradise islands are as much a part of popular culture as sliced bread and pot noodles, and the mere mention of a name such as Bali is enough to evoke a whole host of romantic and colourful associations. Bali's widespread familiarity is due in part to Rogers and Hammerstein's film *South Pacific* (1958), the musical set on a remote tropical island during the Second World War. Unfettered by geographical niceties, Hollywood took the 1930s image of Bali as a paradise island and recast it as 'Bali-Hai', the dream island of American servicemen, and a respite from the terrors of jungle warfare in the Pacific. By combining all the popular images of the South Seas into one, the North American film industry succeeded in creating a 'paradise of paradises' that has proved to be remarkably enduring (Vickers 1989, 3). In a similar vein, *The Road to Bali* (1952), was an amalgam of images of Southeast Asia and the Pacific and starred Bob Hope and Bing Crosby, as well as Dorothy Lamour playing a beautiful princess and wearing her famous sarongs combined with bits and bobs from Thailand and elsewhere in Asia (ibid., 128). There were even a few token scenes of Balinese dance showing small girls dancing beautifully in front of cardboard cut-outs of what the filmmakers imaged a Balinese temple would look like.

At the start of the 21st century the name Bali remains as potent as ever, but has acquired a more sinister association, namely as the setting of the bombings of 2002 and 2005. Despite these problems, the name 'Bali' appears on shops, restaurants, hotels, books, boutiques, handbags and beach ware, in fact on almost anything that is considered tropical, mysterious, sensual, desirable and pleasurable, including a Indonesian made beer called *Bali Hai*. The name Bali continues to be associated with the rich and glamorous, not least because of wrangles between Mick Jagger and Jerry Hall in 1999 concerning the legitimacy of their Balinese marriage a decade earlier and Jagger's financial obligations to Hall in the event of separation (Norman 1999, 6).

Hutt discusses the phenomenon of cultures looking at one another through the distorting lens of mythology with reference to Shangri-La, which like Bali, has become a prominent brand name worldwide representing a version of utopia, particularly in the context of tourism (Hutt 1996, 49; Boon 1977, 2). So strong are these associations that a lecturer at London Metropolitan University once mistook the name BALI on a seminar title as an acronym for the British Association of Leisure Industries, an organisation that does not yet exist. In Bali itself the acronym of *BAnyak LIbur* (*banyak* = many; *libur* = free time) is popularly used in the sense of a lot of holidays being taken or days off to conduct rituals or a place with an unlimited option for the Balinese and/or outsiders to enjoy their holidays. This chapter explores the changing image of the island of Bali from early exploratory and colonial encounters, through touristic imaginings and national awakenings to the name's ultimate hyper-real (Eco 1986) and simulacrum-like (Baudrillard 1983) detachment from the place whence it came and follows a route charted by others (for example Boon 1977; Vickers,1989; Picard, 1993).

In the early 21st century Bali finds itself at a crossroads, and, like many long established destinations, is having to re-evaluate the way it promotes itself, an especially pressing issue in this case because of the terrorist attacks and the threat of more to come. Bali occupies a special place in the development of international tourism, but in order to appreciate why this should be the case it is necessary to look briefly at the origins of this island's pervasive tourist mythology. Bali's tourist industry has colonial origins, but as the work on Boon (1977) and Vickers (1989) has shown, tourism in Bali did not emerge fully-fledged from the colonial encounter.

Like the many societies Bali is being influenced by a variety of globalizing forces, such as world capital markets, the international labour system, the interaction of nation states and the international military order. Globalization is characterized by the intensification of social relations across the world, linking distant localities so that experiences in one location are influenced by events taking place far away in another and vice versa (Giddens 1991, 64). Different analysts have associated the growth of globalization with different periods; Wallerstein (1976), for example, traces it back to the sixteenth century, whereas Giddens links the first truly global experience to the twentieth. Neither perspective is totally incompatible with this discussion, though it is helpful to link the onset of global social relations in Bali with the period of 'high colonialism', roughly 1870 to 1940 (Scholte 1997, 24). Despite attempts by various interest groups to modify Bali's image, the tropes used by international tourism predominate in marketing and though research on tourist brochures is relatively new and the coverage patchy, some common themes have been identified (Dann, 1996, 61). These global images of Bali are not necessarily closely linked to the original Bali and often exist independently of the island since the symbol of 'Bali' has come to stand for almost anything exotic and leisurely.

Early Global Images

In his work on 'imagined communities' and modern nationalisms, Anderson (1983) has described the psychological impact on Europeans of encounters in the 15th and 16th centuries with elaborate and ancient cultures. The hitherto assumed centrality of Christian culture was challenged by the existence of societies in Asia that had developed independently of Europe. Bali was but one of a number of places that captured the Western imagination as the reports of seaborne expeditions to the East began to be circulated in Europe.

The initial Western contacts with Bali were fleeting and transitory since the early Portuguese explorers, who reached Malacca in 1509 and the Moluccas in 1511 by-passed the island in their eagerness to acquire spices and converts to Christianity. Magellan's expedition may have caught sight of the island, though nobody went ashore, and Fernando Mendez Pinto, the Portuguese navigator may have visited in 1546, though the evidence is unclear. Mention is made of the island under various names – Boly, Bale, Bally – on early charts; Sir Francis Drake called briefly in

1580 and possibly Sir Thomas Cavendish five years later. The Portuguese, however, seem to have been the first to consider the island's commercial possibilities and the Malacca government fitted out a ship to visit Bali in 1585, but it foundered on a reef off Bukit with great loss of life. Those who survived were impressed into the service of the Dewa Agung, the ruler of Gelgel, who provided them with wives and homes but refused to allow them to return to Malacca. At this time Gelgel was the leading court on Bali, having welded together the various Balinese principalities into one domain around 1550, though it was not long lasting. The kingdom went into decline during the reign of the grandson of the founder, whose successor abandoned the *keraton*, palace, of Gelgel, thought to be under a curse, and built a new one in nearby Klungkung. The Dewa Agung was to be less significant than some of his supposed vassals and as Gelgel declined there emerged around a dozen independent kingdoms, of which eight survive as modern administrative districts or regencies (*kabupaten*).

In 1597 Cornelis de Houtman's expedition, comprising three ships, straggled into Balinese waters between 25 December and 27 January, anchoring initially off Kuta and Jembrana, and then possibly later at Padangbai. De Houtman's official record and a detailed personal letter by one of the captains provided Westerners with their first substantial body of knowledge about the island. Cornelis de Houtman was so taken by the beauty and wealth of the island that he named in *Junck Holland* (Young Holland), a description so misleading that later Dutchman were given to thinking that by introducing Dutch civilization and commerce, they were guiding the island towards its manifest destiny (Hanna 1976, 9). A party of de Houtman's crew made a trip to Gelgel, where they described the Dewa Agung as living in a vast palace in a walled town with his harem of 200 wives and his troupe of dwarfs. When the ruler ventured out of his palace he was accompanied by scores of his subjects bearing lances and banners, and he was either borne on a palanquin or rode in a cart drawn by two white oxen. The king took an interest in the Dutch gifts and trade goods, but what particularly caught his attention was a chart with a globe depicted in one corner and asked the Dutch to bring him one on their next trip. The king demanded that his visitors should give him a geography lesson, but on commencing with the islands of Southeast Asia the Balinese ruler was disappointed to find that Bali was so small. The lesson progressed on to the territories of the 'Great Turk', which impressed the Balinese, and then on to continental Europe. When asked about which was larger, China or Holland, Arnoudt Lintgens, the highest-ranking member of the shore party, traced the boundaries of the Netherlands in such cunning fashion that it included Scandinavia, Austria and a substantial chunk of Imperial Russia. Cornelis de Houtman seems to have gone ashore only once, while the Balinese hostages that he had taken were being returned, shortly before departing from the island on 20 February (Hanna 1976, 11).

Back in Europe there was a demand for accounts of voyages to distant and exotic destinations, and published versions of de Houtman's log began to be circulated, usually accompanied with illustrations that were designed to appeal to popular taste and were only thinly rooted in Balinese reality (Hulsius 1598). There is, for example, a print of the Balinese king, shaded by a parasol, and flanked by courtiers dressed in sarongs and bearing what appear to be Southeast Asian weapons. The king is shown riding in a wagon pulled by white buffaloes, though the royal conveyance owes much to European inspiration and was, in fact, based on a Dutch agricultural vehicle known as a *bolderwagon* (Boon 1977,17).

Bali's Hindu religion also fired the popular imagination, especially the practice of widow burning or suttee (*satya* in Balinese). European images may contain information drawn from Southeast Asian sources, but often incorporate stereotypes derived from India, particularly Portuguese Goa (ibid.). In order to illustrate the custom of suttee in Bali, the engravers simply referred to the nearest known equivalent. Boon argues that the original engravings of Bali are worth exploring with reference to Gombrich's notion of adapted stereotypes (Gombrich, 1969). It mattered little that the circumstances in which the women were burned were not identical, the point was that the images, albeit Indianized, set Bali apart from the societies with which European readers were likely to be familiar. European fascination with Bali's Hinduism became inexorably linked to the practice of suttee, which remained a popular motif until the early twentieth century. In European eyes Bali was an extension of India; king honouring and familiar on one hand, wife burning and exotic on the other. Suttee continued to be performed during the *puputan* struggles that accompanied the Dutch annexation of Bali and many women chose this way of death following the cremation of King of Badung.

Bali's relations with the Dutch from the 17th to the 18th centuries were also overshadowed by slavery. Balinese women, in particular, were much in demand as concubines and cooks: as Hindus they were able to prepare the pork dishes that European and Chinese merchants in Batavia (now Jakarta) relished, but which Javanese Muslims abhorred. Balinese rulers had recourse to various ways for procuring the desired number of slaves (Boon 1977, 28, 68). A Balinese who became indebted, for example, often through gambling, risked becoming enslaved and being sold off to foreign traders. Thus the Balinese were seen as a partly civilized people, ruled by cruel despots who sold their subjects to finance their ambitions and decadent lifestyles. Bali's image did not, however, remain static and was revised from time to time in response to political expediency and changing intellectual fashions; popular Orientalism also 'attained a vogue of considerable intensity' in the late 18th and 19th centuries (Said 1978, 119). One of the most noteworthy shifts occurred during the Napoleonic Wars when Sir Thomas Stamford Raffles served as the Governor General of Java (1811–16) as result of a British agreement with the Dutch government in exile. In the manner of the Orientalists described by Said (1978, 125), the antiquarians of the day were more inclined to emphasize the East's noble and ancient past than its degenerate present. For Raffles, Bali was the heir to the glorious civilization that had flourished in Java before the rise of Islam, '...a commentary on the ancient condition of the natives of Java' (Raffles 1817, 2). The Dutch were later to foster the idea of Bali as a 'living museum' of Java's Hindu heritage when they opened the island for tourism in the early 20th century (Picard 1993, 74).

Bali's First Tourists

It was undoubtedly a combined government and businesses initiative that led to the development of commercial tourism in Bali, but one should not lose sight of the fact that the island had long proved attractive to Indonesian islanders, notably holy men and poets from neighboring Java. By as early as the 9th and 10th centuries AD, Javanese holy men had begun visiting the island in order to become hermits, and this seems to have continued for some time. There is also an intriguing document dating from the 16th century that poetically describes the wanderings of a Sundanese mystic from West Java who came to Bali in search of a quiet place to practice yoga. In his lengthy travelogue, Bhujangga Manik, makes the astonishing and eerily prescient revelation that Bali was no longer quiet enough for meditation (Teeuw 1998), but quite why this should have been so remains unclear. Interestingly, meditation has proved to be remarkably enduring attraction with some hotels, such as the Legian Paradiso Hotel offering special deals for those with spiritual leanings; Bali's caves also continue to be much sought after for their tranquility and mystical associations, a favoured spot being a grotto on the northeast coast of the off shore island of Nusa Penida.

The island's first tourist of the modern era could be said to be H. van Kol, a member of the Dutch parliament, since he visited Bali in 1902 at his own expense and not as part of an official assignment. He traveled extensively in the Netherlands East Indies seemingly as a much for his own pleasure as any profit, though he had an ulterior motive. As a member of the 'Second Chamber' of the Dutch Estates General he endeavored to inform his colleagues and constituents about conditions in the Netherlands East Indies, and to exert an influence on colonial policy (Hanna 1976, 91). He was not an inexperienced new comer, having served as a civil engineer for the colonial government in the 1880s, and was secondarily a journalist with a desire to explore lesser-known regions, interview key individuals and investigate recent events. He was also something of a scholar, reading official and unofficial Dutch reports, as well as a junketing parliamentary fact finder who would use his position to bypass Dutch bureaucratic obstructions and the indifference of Balinese rulers (ibid.). The outcome of his investigations was a travel book that would be both informative and commercially viable.

The book in question was van Kol's *Uit Onze Kolonien* (Out of Our Colonies), a tome of 826 pages, including 123 devoted to Bali, which was published in Leiden in 1902. Like so many Westerners before and since, van Kol was smitten with Bali from the outset, revelling in its handsome and animated people, and the splendid seacoast and skyline of southwest Bali. It may have helped that he arrived in July during the dry monsoon when there is little cloud cover to obscure Bali's soaring mountains, notably Gunung Agung. Accompanied by the Dutch *Controleur*, Van Kol was a guest of the Raja of Karangasem and settled down to accumulate detailed knowledge on matters of religion, art, agriculture and other aspects of Balinese life, not forgetting to report on issues of ongoing concern to the Dutch such as slavery, suttee, cremation, taxation and public administration (Hanna 1976, 93).

Van Kol was to visit other parts of Bali, even areas that were reputedly more scenic or cultured than Karangasem, but none surpassed his introductory exposure to southwest Bali. He wound up his visit generally convinced of the efficacy of colonial rule and the benefits derived from it by the Balinese, writing that '…there is great and noble work to be done, and hail to the Dutch if we proceed with this beautiful task in a spirit of dedication and selflessness!' (ibid., 98–99). When he returned in 1910, however, to resurvey the situation he was saddened to report that the Dutch had created serious problems and exhorted his countrymen to be guided by humanitarian and not selfish motives and to work with the Balinese for the benefit of the people.

Emergence of the Bali Trope

In the early years of tourism development, Bali was little more than an appendage to much better known Java. The promotional images were somewhat crudely put together, which Vickers has characterised as the '…era of bare breasts, the most obvious of Bali's attractions' (Vickers 1989, 98). The most evocative images for many tourists – and indeed the source of attraction for some of the artists who settled on the island – were those of Gregor Krause (1883–1960) in his famous book entitled simply *Bali*. This photographic essay featured the rituals, people and landscapes, notably the rice terraces, of Bali and continues to be sold in bookstores under the name, *Bali 1912*. Trained as a medical doctor in Germany, he found employment with the Netherlands Indies army, and served in Bali in the central Balinese kingdom of Bangli, where he became besotted with the island. At the time Europeans in particular were keen to visit far off lands to put the ravages and despoliation of the First World War behind them and find solace in unsullied societies. Krause's book fitted the bill and had everything that those seeking a pristine Eden away from a decadent and destructive Europe could wish for.

It opens with high minded words about the unity of man and nature, and argues that Bali's Hindus had managed to attain such a harmony through their religion and society, putting this unity into practice in their daily lives (ibid., 100). In a photographic study a peasant is shown ploughing his fields and worshipping and worshipping the rice goddess, Dewi Sri, as if to say that the Balinese may manage and cultivate the land, but it remains the possession of the Gods. Krause attributed this outlook to the ongoing influence of Majapahit, and the benevolent rule of the descendents of these colonists from Java, reiterating what had by now become a trope, and was rapidly turning into a tourist stereotype. For Krause this aristocracy was simply a veneer over the 'real' Bali and thus the demise of the ruling houses in the bloody *puputans* was of little consequence to the farmer tilling the land since the gods had willed it, regrettable but inevitable. By wishing away the massacres as being somewhat removed from everyday Balinese experience, Krause combined a naked apology for colonialism with the timelessness of tourist exotica. And of course there were the ubiquitous bare breasts, which in Krause's case were kept firm and beautifully formed by the powerful chest muscles of the women who tilled the land.

Over time Bali became one of the most romantic stops on the tourist itinerary, underpinned by a host of new books, articles, photographic studies and films. Like the explorers and traders before them, the longer staying tourists and more permanent residents were keen to rush into print to let the rest of the world know about Bali's magic. They found a ready outlet for their enthusiastic writings in the many new travel magazines, such as the Dutch-sponsored *Inter-Ocean*, as well as the widely distributed monthly glossies that often published accounts of exotic destinations, most notably *National Geographic*.

Island of Artists

An image of Bali that has proved remarkably resilient began to appear in the late 1920s and crystallised around one man, Walter Spies, and his social set. A multiple expatriate of Russian-German origins, Spies had lived in the thriving artistic community of Dresden before coming to the Netherlands East Indies. He was working for the Sultan of Yogyakarta in Java, where he was studying gamelan, when he met Cokorda Gde Raka Sukawati, the brother of the King of Ubud, who invited Spies to come to Bali. Arriving with a piano, a German bicycle, and a butterfly net, Spies initially made his living from catching butterflies, which he sent in gold leaf boxes to museums in Europe and elsewhere (Hilbery 1983, 21). At first he lived in the courtyard of Puri Saren, but later rented a piece of land in Campuhan (Tjampuhan), where he built his Bali-style studio cum residence cum hotel. This two-storied wooden building with a thatched roof still exists and is one of a dozen or so holiday villas of Hotel Tjampuhan and Spa, which is owned by Cokorda Gde Agung Sukawati's family. He was joined in Ubud in 1929 by Bonnet, the Dutch painter, who became the prime mover in creating the Pita Maha art society along with Spies, Cokorde Gde Agung Sukawati, Gusti Nyoman Lempad, and Cokorda Gde Rai, with a membership of about 125 people from various villages.

Highly competent in several disciplines – painting, languages, music and anthropology – Spies was a somewhat reclusive, especially when painting, individual though he was well liked among both the expatriate community and the Balinese. As a homosexual he came in search of a '…paradise away from the strict mores of Europe, and believed that he had found it in Bali' (Vickers, 1989, 106). On Bali sexual relations between men were not condemned at the time, and Spies's homosexuality may not necessarily have added to his sensitivity, but it appears to have contributed to his openness in his relations with the Balinese, especially the men. Vickers has raised the interesting possibility that Bali's image as a homosexual paradise was one of the unintended consequence on the earlier focus on the beauty of Balinese women, and that the men also came to be viewed in a similar light, and to a certain extent passive. When Spies was tried for having sexual relations with an underage male in 1939, Margaret Mead, one of the most famous anthropologists of the period, spoke in his defence, talking about his '…continuing light involvement with Balinese male youth' (ibid.) and drawing attention to the differences in reckoning age between the Balinese and the Dutch. Her appeals fell on deaf ears and Spies was jailed for his 'crimes'.

Like Krause before him, Spies was drawn to the life of the island's peasantry, seen as embodying the genuine spirit of Bali. He enquired into their legends, music, customs and dances, and took numerous photographs that were later to serve as valuable resource for the illustrations for his book with Beryl de Zoete, *Dance and Drama in Bali*. He painted the peasants with their livestock and the legendary figures set against glimmering and multi layered backdrops, an approach that paid homage to the Russian folk art style of Chagall and the French *dounier* (Sunday painter), Rousseau. Parallels can be drawn between the romanticised folk of Spies's paintings and the 19th century Dutch conception of the noble egalitarian figures of the 'village republic'. Spies also investigated Bali's dark side, notably through his contribution to the film *Island of Demons*, made by Victor Baron von Plessen and Dr Dahlseim. These two members of the German intellectual elite knew little about Bali were thus reliant on Spies, giving him the opportunity to express his own vision of the island (Vickers, 1989,107).

Spies's paintings could command such a high price – 3,000 guilders for one of them – that he was able to live off the proceeds for a year (Hilbery, 1983, 22). As can be seen in the 1999 catalogue of Christie's Singapore, paintings by Spies remain valuable, lot 846 having a guide price of US$ 420,000 to US$520,000. As Spies's fame spread he began to attract the attention of the rich and famous such as Charlie Chaplin, Noél Coward and Barbara Hutton, who visited him at Campuhan, often as paying guests, and they also helped to make Bali famous. Spies also had links with officialdom and in his autobiography, Cokorde Gde Agung Sukawati mentions regular visits by Bruyn, the head of the tourism agency in Singaraja, as well as Dr Roelof Goris, the Director of the manuscript library Gedong Kirtya, which was also in Singaraja. Perhaps one of Spies's most celebrated guests was the novelist, Vicki Baum, who wrote *A Tale From Bali*, in which she described an old Dutchman who had been living in Bali since the end of the 19th century and had been the source of her information on Balinese culture and history. The notes that he supposedly gave her were to lend authenticity to her story, but in fact the Dutchman was merely a thin disguise for her real informant, Walter Spies (Vickers, 1989, 110).

With regard to Bali's image, Spies's also exerted a strong influence on the Mexican cartoonist and author, Miguel Covarrubias, whose *Island of Bali* (1937) has outlasted most other travel books to remain the key introductory text for visitors to the island (ibid., 114). The presence of internationally renowned figures such as Covarrubias (1904–57), who at the time of his death had become a national hero in his own country, indicates how cosmopolitan Bali's artistic and intellectual society had become. This image was officially endorsed through the Colonial Exhibition in Paris in 1931, when the European powers vied with one another to display the cultural wonders of their colonies. Bali was the centre piece of the contribution from the Netherlands East Indies and Spies was called upon to advise on the exhibition; he also worked with the distinguished government archaeologist and philologist, Dr Goris, on the accompanying book, which was illustrated with Spies's own photographs.

Spies encouraged his visitors to discover the real 'folk' of Bali, to see the '… relatively untouched native life' as Mead characterised it (Howard 1984, 190), a shift in focus away from the earlier Dutch scholarly interest in Balinese texts and court culture. The underlying assumption was that the 'folk' were artistic, and the 'everyone is an artist' theme reverberates through the literature of the period. The island's leitmotif, the female dancer, can be seen on maps and atlases of the period showing the products of the Dutch colonies, has remained on tourism promotional material ever since. For some, however, the focus on the island's ritual and artistic traditions was overdone, as illustrated by the sceptical and somewhat patronising tone adopted in a poetic address by Noel Coward to Charlie Chaplin:

> As I said this morning to Charlie
> There is *far* too much music in Bali,
> And although as a place it's entrancing,
> There is also a *thought* too much dancing.
> It appears that each Balinese native,
> From the womb to the tomb is creative,
> And although the results are quite clever,
> There is too much artistic endeavour.
>
> (Clune, 1940, 317)

But this seems to have been a minority view – at least in public – and the idea that Bali was some kind of cultural treasure house was to endure and take on a new set of meanings in the independent republic that eventually replaced the Netherlands East Indies after the Japanese occupation of the Second World War, and this was particularly true of the tourism boom years under Suharto (1966–1998).

Tourism and Nationalism

Bali, which had been part of the province of the Lesser Sunda Islands, became a province of the Republic of Indonesia in 1958 with its own governor and two years later the capital in the former Dutch residency of Singaraja (North Bali) moved to Denpasar, which was closer to the airport. The Dutch had hitherto kept modernisation at arm's length, but the nationalists who were now in power wanted to bring the benefits of progress and education to their people, and they reacted against the idea that Bali was a kind of 'museum' society at the service of tourists seeking exotic nostalgia (Picard 1996, 41). Spurred by concerns for morality and decency the colonial authorities had long before ordered Balinese women to cover their breasts in public, but this was only adhered to in Singaraja and Denpasar; the newly empowered Balinese nationalists reimposed these measures. Another measure aimed at dismantling the tourist image that many Balinese found embarrassing was the introduction of a rule that forbade visitors from photographing bare-breasted women. In fact, in 1939 the nationalists had urged the colonial regime to stop tourists from photographing the exposed breasts of Balinese women and to ban the circulation of postcards and pamphlets with images of topless women (ibid.).

This stance presented the nationalists with a dilemma since the island was suffering an economic depression and, despite their radicalism, the nationalists came close to acknowledging that only tourism could restore the island's fortunes. Sukarno, whose mother was Balinese, exhorted his countrymen to look to their own traditions rather than emulate the West, but at the same time the nationalists were involved in the creation of an undivided nation state, which meant either shedding ethnic identities (seen as a hindrance to modernisation) or re-conceptualising them as part of the nation's exemplary past. The nationalists were aware that in order to stimulate tourism, Bali would have to nurture the cultural particularity that had set it apart before the war, but this contradicted the spirit of national integration. Such concerns were not yet a serious problem since the number of international tourism arrivals remained very limited, inhibited by the poor state of the infrastructure, constant political agitation and the xenophobia of the regime (ibid., 42). Accession to Indonesia had, however, opened Bali to a new clientele – the Indonesian governing elite; Sukarno adopted Bali as a holiday retreat, making it an obligatory stop for esteemed guests.

In order to differentiate the new nation as sharply as possible from the former colony and mother country, one might have expected the Indonesian nationalists to draw a sharp distinction between Indonesian and European cultural heritage (Holtzappel 1996, 64). This would have been in accordance with the official view, but in practice the evolution of the Indonesian nation state owed much to European inspiration. The means used to build the new state were somewhat reminiscent of the 'invented tradition' processes described elsewhere by Hobsbawm and Ranger (1983), though the Indonesian nationalists in the fifties intermittently denied any debt to the West. The *puputans*, for example, that had shamed the Dutch earlier in the century were re-formulated as early independence struggles and the Badung 'fight to the death' is commemorated in a large monument in Denpasar, the current one having been erected in the late 1970s. The Badung *puputan* has only been celebrated annually in Denpasar since 1973 and earlier attempts to commemorate the event were probably discouraged in favour of Sukarno's nation building projects and Suharto's early development plans. One distinctive feature of the new state was, however, the adoption of a national philosophy of *Pancasila* (five principles), which was taken from Old Javanese sources, *Kakawin Nagarakertagama* (Holtzappel 1996, 103). What was significant about Pancasila for the Balinese and other religious minorities within Indonesia was that the first principle or *sila*, as symbolised by the star on the national code of arms, guaranteed freedom of worship. Provided one believed in God the religion was not specified and this would later have implications for Bali's tourist image, not least when it was portrayed as a peaceful Hindu enclave during the upheavals of the Asian Crisis.

Sukarno's rule effectively ended when an attempted coup by Indonesia's communists in 1965 failed and sparked off a counter coup and widespread massacres. Sukarno lingered on for a while, but power was eventually seized by General Suharto and what was to become known as the 'New Order' government. One of the new regime's first objectives in the context of development of the ruined national economy was to restore tourism in Bali, which had experienced severe strife, some of it witnessed by holidaymakers. According to Picard, the spectacle of thousands of holidaymakers queuing to visit the island, enabled the new government to claim that it had earned the confidence and respect of the rest of the world (Picard 1993, 95). To a certain extent, the regime's methods were not unlike those of the Dutch earlier in the century; in both cases tourism was seen as an effective means of salvaging each regime's reputation; it was also a pragmatic step in an economy blighted by strife.

The culture of Bali came to be seen as a resource, one of the 'cultural peaks' of the emergent national culture whose function was to facilitate the growth of tourism and foster national pride. Especially significant in this context appears to have been the ideas of Ki Hajar Dewantara who actively supported the development of many regional cultures that would subsequently contribute to the emerging national culture. Together these cultural peaks, *puncak-puncak dan sari-sari kebudayaan daerah*, would lay the foundations of a rich national culture (Nugroho-Heinz 1995, 16–17).

An insight into how Bali is perceived as a regional culture, part of a set of many, can be derived from the way the island is presented in museums, most notably in the open-air village museum that forms the core of the Jakarta leisure complex known as Taman Mini. In order to appreciate why a museum should serve as a showpiece of Indonesian identity, it is worthwhile considering briefly how the nation came into being. The territory today known as Indonesia was created in response to many centuries of European expansion as the borders between the competing colonial powers gradually became established. This process came to a halt during the first decade of the twentieth century when the islands that were not yet fully incorporated into the Netherlands East Indies were finally absorbed. The inhabitants of the vast region, known after independence as The Republic of Indonesia, shared Dutch rule regardless of their ethnic or religious affiliation, creating a kind of negatively defined consciousness. According to Hubinger, Indonesian nationalism was born out of the Herderian brand of nationalism in which the people are the nation's founders (Hubinger 1992, 4).

What is significant is that Bali, in common with the other 27 Indonesian provinces that were represented in Taman Mini at its inauguration, were expected to provide nuances of colour (*aneka warna*) to the national culture (Picard, 1993, 92). The point being that the province and not the ethnic group had become the source of culture; Bali was not a state within a state, but a regional culture, *kebudayaan daerah*, an integral part of the cultural heritage of Indonesia. What should not be overlooked, however, is that Taman Mini also serves a very useful purpose, in that it provides both overseas visitors and Indonesians with an introduction to the complexity of Indonesia as a nation. Taman Mini is also a genuinely popular recreation area, providing open spaces and variety within the highly urbanised region of Jabotabek, as Jakarta and its neighbouring towns are collectively known.

Under Suharto Bali was expected to take its place in the new national order, part of a set of many, but the end of the oil bonanza in 1983 forced the government to take a new look at its tourism policy. Bali was far too well known abroad, often better known than Indonesia itself as government officials often cheerfully admitted, to be allowed to drift into obscurity. The Ministry of Foreign Affairs advanced a policy of cultural diplomacy, which rapidly became known as *gamelan* diplomacy after the Indonesian gong orchestras that travelled the world in the 1980s and early 1990s advertising Indonesia's cultural riches. According to Picard the aim was to utilise the nation's cultural riches to promote Indonesia abroad as a country of 'high culture'. In accordance with their policy, the Balinese dance troupes that were sent on tours were expected to serve as artistic missions (*misi kesenian*) to simultaneously develop international tourism and promote Indonesia's cultural image. One of the most significant attempts at cultural diplomacy took place in the 1980s when Indonesia and the USA agreed to hold a promotional event called KIAS, *Kebudayaan Indonesia di Amerika Serikat*, aimed at raising American awareness of Indonesian culture, including 'regional cultures' such as Bali. As a recognised 'cultural peak' Bali was supposed to represent Balinese identity on one hand and Indonesian identity on the other, and thus Balinese culture was placed in a similar position with regard to both tourism and Indonesian nationhood (Picard 1993, 94).

The Impact of Image

What studies of tourism such as Picard's suggest is that tourism cannot readily be isolated from many other aspects of culture, and this is particularly the case with places such as Bali which have a long history of tourism. By treating tourism as a solely exterior force analysts risk ignoring how tourism can become part of the local reality: in certain cases tourism and traditional culture have become inseparable. When culture is conceived of as a static entity, lacking the dynamics of change, then the actions, motivations of local participants are overlooked. Artistic styles, performing arts and even changing food habits, can be seen as local attempts to accommodate to – and in many cases to profit from – the cultural experience of tourism. By using more actor-orientated studies we can build up a clearer picture of how people respond to tourism and what tourists actually do. Balinese reactions to global tourism have also been charted through a series of cartoons that have appeared in the local newspaper known as the *Bali Post*, which is published in Indonesian, and in the newly founded Balinese English cartoon magazine, *Bog Bog*.

Bali provides an interesting case study in this respect because the longevity of its tourism industry furnishes us with the opportunity to see how the interface between tourism and traditional culture works over time. A good illustration of how this can be achieved is provided by Annette Sanger's study of a *barong* dance drama in the Balinese village of Singapadu (Sanger 1988, 89–96). Sanger shows how the barong dance dramas, which are important rites that protect the community from malevolent forces, have been adapted to suit the needs of tourists and local villagers in several ways. The village that Sanger studies is off the beaten track, but is usually visited by tourists around three to four times a week. Tourists stay around 70 minutes, allowing them enough time to watch a performance and buy some souvenirs. In the early days spectators used to sit through performances lasting around three hours, but the villagers realised that this increasingly had less appeal. A new dance programme, which included slapstick sequences to help transcend language barriers, was therefore devised by one of the best dancers in Singapadu. Performances were also reduced in length and female dancers were introduced to play some of the female roles that had hitherto been played by men. The villagers also modified some of the kris dances, appreciating that some tourists disliked seeing dancers come out of trance by eating live chickens.

The villagers do not, however, see these changes in terms of cultural denigration and justify the commercial use of dance dramas in the following ways. First, the barong is still treated respectfully and the correct rituals are performed. Second, the oldest and most sacred barong costume is not used for commercial purposes and is thus not desecrated. Third, the villagers value the revenue derived from tourism and, since the remuneration goes to the community as a whole, then no individual is guilty of greed. Fourth, the barong likes to dance, whatever the circumstances. Fifth, performances reinforce community solidarity, and villagers maintain that they miss having the opportunity to come together during quiet periods in the tourist cycle. Sixth, the performers argue that they never compromise on technical and aesthetic standards, and moreover tourist performances provide them with an opportunity to practice. It would appear that local traditions have been adapted to the needs of tourism in subtle ways, suggesting that a re-think of some of the more elementary observations of the impact of tourism is long overdue. Goffman's celebrated distinction between 'front stage' and 'back stage' would appear to have analytical value here (1958).

What is also significant is that debates concerning tourism development have from time to time reached the public domain, allowing us to see how the Balinese have attempted to mould the industry to their needs. Boon, for example, has discussed a scandal involving a seaside shrine that lay in the path of a proposed tourism project. There was resistance on the part of the local people between 1971 and 1972, but the local district head still sent in a demolition team. The team, however, lost its nerve at the last moment and the sacred monument was saved (Boon 1977, 216).

In the late 1980s there were also a series of demonstrations concerning the planned redevelopment of the famous temple of Tanah Lot on the coast. The tourism authorities wanted to upgrade the facilities and prevent accidents by concreting the slippery path used by tourists visiting the historic site. In 1989 the local villagers and activists protested arguing that the plans would spoil the natural beauty and ecology of the coastline. These activities should not simply be interpreted as anti-tourism, not least because the protestors made it clear that they appreciated their contribution made by tourism to the economy. A potential solution would have been to leave the path as it was and to provide better interpretation for the tourists, warning them about the access problem.

Bali as a Brand

Despite attempts by various interest groups (for example inter-war artists, successive Indonesian governments) to modify Bali's image, the tropes used by international tourism predominate in the market place. These global images of Bali are not necessarily closely linked to the original Bali and often exist independently of the island. The leitmotif of 'Bali' has come to stand for almost anything vaguely equatorial and exotic, and the proliferation of the 'Bali' brand name continues apace. There are also signs that other tropical islands are being subjected to a similar kind of treatment, not least the new would-be Bali of neighboring Lombok. In 1999, for example, a ubiquitous Thai and South-East Asian restaurant known as 'Lombok' opened in Herne Hill in South London, close to the residence of the British author of this book.

It is not possible at this stage to compare and contrast the various forms of image creation discussed here with those produced in other contexts within international tourism, though the categories of paradise proposed by Dann (1996b, 222) would be worth analyzing with reference to a number of different destinations. Some clues as to how tourists understand these images are provided in a song called *I've Been to Bali Too*, which was written by John Schuman for his group Redgums. It announced the arrival of Bali as Australia's Costa del Sol and also seems to have become a catchphrase since part of the song cheekily compares Bali with other world destinations favoured by Australians in the 1980s:

You can't trick me
'cause I've been to Bali too
You've been to Paris and you've
Been to Boston
You've been to Fiji and you've
Been to London
But you can't impress me
'cause I've been to Bali too

(Ellis 2004, 31)

The identity issue is likely to become more pressing as the media begins to adopt a more critical stance towards tourism, though not necessarily underpinned by sustained research. A BBC2 programme entitled 'The Shape of the World: Global Tourism', for example, argues that Hawaii's image as a tropical paradise is resented by its people who believe that tourists are exposed to an artificial version of their culture.

Our picture of tourism development in Bali is far from complete, but what the above accounts – and the previous two chapters – suggest is that the Balinese reaction to tourism is at least in part flexible and pragmatic. Picard argues that in becoming the brand image by which Bali is differentiated in the international marketplace, culture has become a defining identity marker. Balinese culture is thus simultaneously what defines Bali from the outside world and what connects the island to it (Picard 1996, 197). Touristic culture blurs the division between how tourists identify the Balinese and how the Balinese identify themselves. Encouraged to preserve and promote their cultural identity with reference to the outside world's view of Bali, the Balinese appear to re-confirm their Balinese qualities by recourse to images purveyed by global tourism. The cultural image that arises reflects interaction with various interlocutors, including the overlapping networks of global tourism, the media and the Indonesian state, to mention a few. The outcome is neither solely global nor local, but is a reflection of the interaction or collision between them all, as it will be discussed in the next chapter dealing with a selection of Balinese villages that have been drawn into tourism in various ways.

Chapter 4

Bali's Global Villages

Implicit in the terms words 'village' and 'community' are notions of local social groupings that are self-contained, cohesive and cooperative, while at the same time being old fashioned, static and chauvinistic. These positive and negative connotations, however, seriously misrepresent the complexities of rural societies, especially in Asia where an exaggerated stereotype of the homogenous and unchanging village community has pervaded academic discourse since colonial times (Warren 1991, 214). In the case of Bali it is particularly misleading to treat 'the village' as if it were a concrete manifestation of Wolf's (1957) ideal 'closed corporate community' since the island's villages are neither homogenous nor closely integrated, a far cry from the *dorpsrepubliek* (village republic) image projected on Bali by classical Dutch scholars (Geertz 1961). The idea of the village republic arose out of attempts by Dutch scholars to account for what they perceived to be the anarchic condition of Balinese government. They entertained the view that villagers had a voice in choosing the lord to whom they paid homage and taxes, and often exercised this right by fleeing from unpopular masters (Vickers 1989, 90); a Balinese kingdom, especially one not under Dutch rule, was seen as little more than an agglomeration of villages (Liefrinck 1927, 31–2).

In reality, Balinese villages are socially and economically stratified, are not centrally organised and do not conform to the bounded and cohesive communities implied through conventional discourses on village communities. The socio-economic inequalities among Balinese villagers do not concur with an image of homogeneity and, though it would be misleading to overemphasise these differentials. There clearly are corporate features that underpin collective economic and political action, though they have there own delineated spheres of jurisdiction. At the village level in Bali, the main corporate institutions are: the *desa adat* (customary village), *banjar* (hamlet or ward), *subak* (irrigation association), *pemaksan* (temple congregation), *dadia* (patrilineal descent group) and *sekaa* (voluntary work group) (Warren 1991, 215). These groupings are formally independent, though irrigation society and descent group memberships cut across village boundaries, and connect with one another in a pattern of intersection that defines the Balinese social system. Customary codes specify their field of jurisdiction and there is a hierarchical ordering, in which the needs of the hamlet outrank those of the descent group, and both take precedence over the voluntary work group. Each corporate group has its distinctive oral or written customary codes (*awig-awig*), specifying in detail the rights and obligations of its members. The Balinese institutions that most correspond to the Western notion of a territorial village or community are the customary villages and the hamlets respectively, though the analogue should not be taken too far.

The idea of a 'global village' arose in the 1960s when new communication technologies appeared to be shrinking the world into a village-like community (McLuhan 1968). The later term, globalization, represents the other side of the process whereby villages that were once isolated are progressively drawn into worldwide cultural and economic systems. Balinese villages, notably Ubud, have been promoted as distinctive local products, though their development has since the early 20th century been inseparable from links with foreigners (MacRae 1999, 124). Ubud might be said to be a global village of the latter type, a formerly agrarian village whose fortunes are now intertwined with international tourism despite the political turbulence of the late 1990s and the early 21st century. Despite these changes, the villages have not yet disappeared and as MacRae points out the people of Ubud act globally as one might expect, but think surprisingly local (ibid.) The people of Ubud are not ignorant of the global forces on which their livelihoods depend, but they continue to conceive off their own prosperity, along with other aspects of their lives, in terms of their relationship with Bali-Hindi gods. Similar observations can be made, with some important local caveats, with three entirely separate villages that have hitherto not been discussed much with reference to tourism.

The Mountain Balinese

The so-called 'mountain Balinese' (*Bali Aga*), whose name is derived from an Old Javanese term, *aga*, for mountain, reside in the uplands of central and east Bali. They are also less commonly referred to as the Bali Mula or Kuna, which means 'original' or 'ancient' in Balinese. Because the term 'mountain Balinese' is problematic, but is more ambiguous than 'original' or 'ancient' which are too polarised in their implied values (Reuter 2002, 12–14), the term 'mountain Balinese' is used here. In 1992 the Regent, *Bupati,* of the District of Bangli, Ida Bagus Agung Ladip, issued a letter of intent, *Surat Keputusan Bupati Bangli*, which re-classified one of the mountain Balinese settlements, Penglipuran, as a 'traditional tourist village', thereby turning into an official tourism destination.

What makes this village doubly interesting is that it is also situated in a part of Bali that is closely associated with the anthropological work of Margaret Mead and Gregory Bateson on the psychology of childhood. The connection is with the village of Bayung Gede where Mead and Bateson conducted a substantial portion of their research, and the villagers of Penglipuran claim to be descendants of migrants from that village. Mead and Bateson conducted research in Bayung Gede between June 1936 and June 1937, and then intermittently until February 1938, having moved their base to the *raja*'s palace in Bangli. Neither Mead nor Bateson wrote a monograph on Bayung Gede and most of what they observed about the community has to be gleaned from references in articles, books, and Mead's letters to home and friends (Sullivan 1999, 7). Mead and Bateson were by no means the only social scientists to carry out research in Bali and the majority of studies of the period were written in Dutch, such as V.E. Korn's works on Balinese law and culture, including a study of the Bali Aga village of Tenganan Pagringsingan (1933). Many contemporary Indonesian researchers have either studied at home in Indonesian with access to

English or abroad in Anglophone countries such as the USA and Australia, and Dutch has become increasingly inaccessible for them and it is Mead and Bateson's *Balinese Character* that is more widely known today among the Balinese intelligentsia. It would thus not be unreasonable to assume that even if the Regent himself did not know about Bayung Gede's anthropological associations, then at least someone working on his staff in Bangli did.

In view of the fact that Penglipuran is a comparatively new village by Balinese standards, why was it and not Bayung Gde that turned into a 'traditional tourist village', a surprising oversight when one considers how well known, at least among academics, Mead and Bateson's publication was? Anthropological research, moreover, could be said to be an intellectual version of globalization since it can bring about an intensification of relations, often unintentionally and often through tourism. Anthropologists often attempt to spare the villages that they study from unwanted attention by outsiders through the use of ethnographic pseudonyms, but Bali's villages are so well mapped, especially in the south-central region, that it has become increasingly difficult to do this. Tourism also readily makes use of stereotypes – some of them derived from early anthropological accounts – that influence the way the mountain Balinese are represented in official accounts, guidebooks and other kinds of media. Given that the government's decision to turn Penglipuran into a tourist attraction may have been at least partially influenced by an anthropological consideration. It is worth asking why governments and anthropologists alike appear to share a common interest in these kinds of villages. There are also significant developmental issues to be considered, not least because the mountain Balinese are in a relatively poor position '...to claim a significant stake' in Bali's cultural tourism (Reuter 2002, 16); in fact Tenganan in the Karangasem regency, Penglipuran and Truyan are the only Bali Aga villages to derive significant benefit from tourism (ibid., 18). This raises the sensitive question about whether the history of the development of anthropology as a discipline could be used help raise the island's profile in a positive way, a particularly pressing question in view of Bali's recent crises?

In general, the upland peoples are regarded as the original (*asli*) inhabitants of the Indonesian archipelago and are often seen as not fully integrated members of the world religions – Hinduism, Buddhism and Islam – that flourish in lowland and coastal areas. But the notion that the 'mountain Balinese' are in any sense 'original' is misleading since even though they have much in common with the peoples of eastern Indonesia who were not profoundly influenced by the long established world religions, the Bali Aga were brought into the Hindu orbit just like their lowland neighbours. The introduction of Hinduism in Bali created new forms of ranking through the adoption of ordered descent groups that have some affinities with castes in India. All Balinese are referred to as *caturwangsa*, but the gentry, who bear caste titles and roughly correspond to the *brahmana*, *satria* and *wesia* in the Indian archetype, are known as the *triwangsa* (Eiseman 1989, 32–34; Hobart et al. 1996, 76). The influence of the Hindu-Javanese during and after the Majapahit era in Bali (AD 1343–1528) affected the 'mountain Balinese' less than other islanders to the extent that members of the higher castes often refer to themselves as people of Majapahit in contradistinction to the upland dwellers. Unlike the majority of the islanders, however, the 'mountain Balinese' neither have castes, nor cremate their dead and have thus come to symbolise, if not completely to remain, the non-Hindu element in Balinese society (Boon 1977, 190).

The first anthropologists to try to comprehend the cultural complexities of the archipelago were often struck by the differences between lowland peoples and those of the interior and uplands. Similar observations were also made with regard to mainland Southeast Asia where the court-based societies of the valleys often exercised only limited control over the upland dwellers. A commonly held view was that these cultural differences were due to waves of migration over time with the more sophisticated peoples, known Deutero-Malays, following less-developed ones, so-called Proto-Malays, and settling the lowlands (cf. Keers 1948; Heberer and Lehmann 1950). Such perspectives were taken up by more popular commentators on Balinese culture such as the Mexican artist, Miguel Covarrubias, who resided in Bali during the inter-war years, and saw the Bali Aga as the 'pure descendants' and the 'original Balinese' (Covarrubias 1937, 17) among whom the 'old Indonesian spirit' had remained (ibid.,167).

Among the first to conduct an in-depth study of a Bali Aga village, were Mead and Bateson who accounted for their distinctiveness in terms of a cultural base on to which intrusive elements had been grafted (Mead and Bateson 1942, xiii). A student of Franz Boas, Mead was an advocate of the view that social factors rather than hereditary ones shaped human behaviour, the so-called 'nurture-nature' debate, and her work in Bali is presented in these terms. These anthropologists described the Bali Aga as having no Brahman priestly households, no names for the Hindu gods and 'ceremonially bare' as compared with other Balinese mountain dwellers (Mead and Bateson 1942, xiii). For many mountain Balinese, Bali Aga is a term used by others, which some, particularly around Trunyan on Lake Batur, consider degrading (Hobart et al. 1996, 27). Moreover, not all the people considered to be Bali Aga dwell in the mountains, a case in point being the villagers of Tenganan, who are said to be of divine origin and to be able trace the origins of their social system to a written source, a holy book (Ramseyer 1984, 194). In addition, Tenganan, which has fertile and well-cultivated lands, is not typical of the other Bali Aga villages, though like them it seems to have remained largely outside the Majapahit sphere of influence (Covarrubius 1937, 20; Ramseyer 1984, 195; Sulistyawati 2000).

During the second half of the 20th century terms such as Proto- and Deutero- Malay gradually fell into abeyance and the position on the so-called 'mountain Balinese' had to be substantially modified when it was shown that there were royal edicts, even in the era of the early kingdom, that specifically referred to the Bali Aga or Bali Mula. What became clear was that the Bali Aga had had a long experience of dependence on the courts and temples, and had maintained significant commercial relations with the outside world. It thus became misleading to identify these communities with an original community that retained an old Indonesian identity (Hobart et al. 1996, 26–7). A more plausible explanation was that these people were located early in the Hindu Balinese royal administrative network, but somehow avoided later reforms, which were inaugurated elsewhere under the auspices of Majapahit. Despite the above developments, the notion that people such as the mountain Balinese are some kind of aboriginal population continues to guide Indonesian policy and lives on in popular scientific accounts, often finding its way into official guidebooks (Badung

Tourism Promotion Board 1978, 145). Bill Dalton's bestselling tourist handbook on Indonesia specifically mentions the Proto- and Deutero-Malay issue (1978, 4) and refers to Tenganan as an 'original Balinese settlement' (1978, 205). The fact that these ideas filtered into tourism is unsurprising given the huge popularity of books, such as Covarrubias's *Island of Bali* (1937) that remains in print long after the author's death and continues to espouse such views.

Ritual Domains

The mountain Balinese communities conceive of themselves as being interconnected via series of migrations that shape the cultural landscape of upland Bali through regional networks of alliances between groups of villages, which are referred to locally as *banua*, or 'ritual domains' (Reuter 2002, 25). The subsidiary settlements are said to have emerged from clusters of temporary dwellings in garden plots that were too far removed from the original village for the occupants to return from work on a daily basis. The village of Penglipuran is thus one of a number of downstream garden plot hamlets to be founded by ancestors from Bayung Gede (Reuter 2002, 39) and is now a village in its own right with its own set of temples. The link is, however, commemorated and Penglipuran's inhabitants remain ceremonially obliged to observe major festivals at the temples of Bayung Gede, approximately 20 km to the north. The name 'Penglipuran' may be derived from the words *pangeling* (remember) and *pura* (temple), possibly characterising the villagers as those who remember the temple, perhaps in the sense of being mindful of their religious obligations (Aryantha Soethama 1995, 46).

In accordance with an alternative or possibly linked explanation, the villagers were encouraged to settle lower down the hillside by the ruler of Bangli. The villagers of Bayung Gede were thought to be very strong and this is reflected in the name of the settlement (*bayu* = power, *gede* = great) (Aryantha Soethama 1995, 46). The *raja* needed strong labourers to work on the land and in particular required physically fit people to carry the many tiered funeral towers that were used for royal cremations. The practice of encouraging upland peoples to re-settle in more accessible areas was moreover a means by which the ruler could exercise more control over or have more contact with his subjects. In accordance with this perspective, the name of the village may also be derived from the word *lipur* (refreshing), a reference to the use of the cooler uplands by the *raja* for recreation (Aryantha Soethama 1995, 47). The policy of encouraging upland peoples to re-settle in more accessible locations continued in post-colonial Indonesia and well into the 1990s and there are many cases of highlanders being asked to move to sites where they could in theory have more access to the benefits of development, though this was often accompanied by more government intrusion. The stark material consequences of remaining apart from mainstream lowland society have been noted among the Dou Donggo of Bima, Sumbawa by Just (2001, 3), but the advantages of complying with government inducements are not always apparent. Even when complying with the government's inducements and moving to an approved location, as was the case with the highland Ngadha of Flores, development does not necessarily follow automatically (Cole 2001).

Like their relatives in Bayung Gede, the villagers of Penglipuran do not have a caste hierarchy, though they are Hindus and are regarded as *jaba* by the lowland nobility. Historically the villagers were expected to abide by royal edicts, but were permitted to retain autonomy over internal village affairs. The villagers have an egalitarian ethos with an emphasis on communal obligations rather than the individual accumulation of goods. The villagers are divided into descent groups, known *soroh*, and each one traces descent from a common founding father, either real or fictional, and a particular place. Unlike the descent groups of southern Bali, the *soroh* of Penglipuran are not strictly ranked and are more united, partly because of a shared sense of common origin; membership of the group is expressed through the preparation and performance of festivals in the temple of the common ancestor. The descent groups are divided into households comprising an extended family of three or more generations living in brick walled compounds lining the central street. Access to each compound is via a short flight of steps and an arched entrance, which is closed and barred with wooden doors at night. Despite Penglipuran's long association with a Balinese kingdom, it has retained a distinctively Bali Aga form of social organisation and it is likely that its traditions were once more widespread over the Bangli regency as a whole (Reuter 2002, 191). In the eyes of the Balinese intelligentsia, and in comparison with south Balinese settlements, Penglipuran is a very traditional village.

'Traditional Tourist Village'

The title 'traditional tourist village' sounds strange in English, but represents a reasonable translation of the Indonesian terminology: *Desa* (village) *Wisata* (tourism) *Tradisional* (traditional). The designation appears to have been adopted by the regent and his advisors in order to conserve what they regarded as the unique heritage of the village combined with a desire to harness more of tourism's development potential in Bangli by creating a distinctive attraction. This attempt to conserve a living village community within the context of tourism may be of partly European inspiration (Aryantha Soethama 1995, 55), and there are examples of village-based heritage projects incorporating European approaches elsewhere in Indonesia (Hitchcock 1998, 128).

The 'traditional tourist village' concept could easily appear to be a relic of the Suharto era with its commitment to the village as the exemplar of the Indonesian way of life, an outlook that is detectable in countless projects throughout the archipelago, starting with Jakarta's multi-cultural showpiece, Taman Mini, in 1975. Villages, often purpose built like Taman Mini, were either created from scratch or modified, usually spruced up, to support the twin projects of nation-building and tourism development. Tourism was encouraged to increase foreign revenue, enhance the nation's international status and foster domestic brotherhood (Suharto 1975, 156–7). Penglipuran and the other Balinese settlements that were designated as 'tourist villages' came into being towards the end of the Suharto period when the issue was not so much nation building, but cultural conservation in order to service tourism. Alarmed by the notion that tourists might not like a developed and thoroughly

Indonesianized Bali – the two missions usually went hand in hand in Suharto's time – the authorities set about conserving what they saw as 'traditional', usually opting for the more picturesque villages.

After Penglipuran had become a 'traditional tourist village', a special gateway was built and a sign was erected proclaiming its special status in both Indonesian and English. From this point onwards entrance to the village was by ticket only, though local Balinese were exempted, and a series of conservation measures were introduced. The restrictions imposed by the regional authority were almost exclusively concerned with material culture, namely handicrafts, roads, verges, gateways and dwellings (i.e. tangible heritage). The Regent ruled that the buildings facing the main street in Penglipuran were kept intact and were to be repaired and maintained using traditional materials only, namely bricks and mortar, timber and thatch. The villagers were also forbidden from opening tourism related businesses, such as food stalls and souvenir shops, in the conservation zone, though these rules were never strictly enforced and small businesses catering to tourists quickly appeared. The local government also fostered a sense of civic pride and encouraged the villagers to tidy up the main street and keep it clean at all times. These changes were enacted through the village assembly to which members were summoned by beating on a wooden drum, *kulkul*; failure to send a representative resulted in a small fine.

Penglipuran represents a relatively new kind of rural village in Bali and is moreover one of the few whose entry tourism was specifically encouraged by government edict. In his account of the International Conference on Cultural Tourism, which was held in Yogyakarta, Java in 1992, Picard provides insights into what motivated this decree. In response to widespread concern of the perceived growing gap between what tourists are supposedly seeking in Bali – 'unspoiled landscapes and authentic culture' – and what they find the authorities wished to preserve especially picturesque villages in their traditional condition for tourism (Picard 1996, 189). At the conference the Governor of Bali, Ida Bagus Oka, gave a paper bearing the title 'Universal Tourism: Enriching or Degrading Culture?' which advanced the concept of 'village tourism' (*pariwisata pedesaan*) and publicly announced the decision to create three 'tourist villages': Penglipuran (Bangli), Sebatu (Gianyar) and Jatiluwih (Tabanan) (ibid.). Picard cautions against being fooled by the rhetoric of public officials of the Suharto era, especially when it is known that many were self-serving and not especially concerned about Bali's sustainability as a tourism destination.

The houses flanking the road are constructed in the Balinese manner on raised stone and cement platforms with rendered brick walls. Some dwellings have thatched *alang-alang* (*imperata*) grass roofs, but the majority are either tiled or covered in corrugated iron. The majority of compounds have a small timber and bamboo building with a sharply pitched roof of the type photographed by Walter Spies and Beryl de Zoete in the 1930s (Hitchcock and Norris 1995, figs 59–60). These dwellings sit atop mud and stone platforms and comprise a single room, which is used for cooking, storage and sleeping. Each family's most senior married couple usually resides in these houses, and when they die it is the custom for the next oldest pair to move in. Some of these simpler dwellings appear not to be inhabited suggesting that the most traditional buildings are no longer attractive for at least some elderly villagers. As one moves away from the central street towards the rear of each compound more modern looking buildings made of concrete blocks become apparent, some of which serve as workshops for producing hand-carved wooden souvenirs. The more wealthy households have televisions and other electrical appliances, and some own expensive four-wheel drive vehicles.

During the day the majority of men are absent working in the rain-fed fields well away from the village and thus it is the women, some of whom speak the visitor's languages (i.e. English, German, French, Japanese), who have most contact with tourists and who try to lure visitors into their compounds to patronise their souvenir stalls. The women also rent out rooms to tourists as 'homestays', and are permitted to do so by the regency provided they do not alter the character of the buildings. Women are also heavily involved in assembling and painting souvenirs, though much of the carving is still undertaken by men. Over half (56.52%) the working population are involved in agriculture and almost a quarter (22.22%) work for the state. The rest are employed as traders (3.86%), handicraft makers (5.80%), and skilled tradesmen (8.70%), whereas 2.907% leave the village to work for the armed forces, ABRI (Aryantha Soethama 1995, 54). Some of the villagers are connected to global tourism in an unexpected way given their mountainous locations, and work for an American-run cruise ship company.

Changes to customary law may only be enacted providing that they have been subjected to scrutiny and discussion among the village's council of elders, and there are examples of adaptations to tourism. It was, for instance, the council of elders that decided to close off the village temple to tourists in the 1990s on the grounds that foreign women might inadvertently enter the shrine while menstruating without realising that they were committing an offence under customary law. By 2002 this decision appeared to have been revoked and it was once again possible for foreigners to enter the temple, though the inner sanctum was still barred to women.

Signs in English warning visitors about the need to respect local customs have been posted, but aside from a billboard, there are scarcely any other indications that the village is still living in accordance with customary law, albeit a traditional code that can be modified. Visitors may be able to appreciate the tangible aspects of the village's traditions, but, in the absence of much interpretation, they are given very little insight into the village's intangible heritage. There is, for example, no indication that in matters of customary law the council of elders still rules supreme and that they cannot in theory be over-ruled in these matters by the representative of the Indonesian state, the village headman. Indeed, apart from the sign by the entrance written in English and Indonesian, there are few obvious symbols of the presence of the state that would be obvious to tourists. Perhaps the most striking feature that would be intelligible to speakers of Indonesian is the location of the *Camat*'s (sub-district head) office close to the road leading to the village where government rituals, such as flag raising, are performed on a daily basis. Researchers and businessmen are expected as a courtesy to check in with the office before entering the village, but tourists are unlikely to be aware of its presence.

Mead, Bateson and Tourism

Penglipuran was subjected to a top-down and clearly essentialist view of what a traditional village should be like; this did not automatically result in *de novo* creation and a complete break with prior meanings and representations. By comparing photographs of the village with architectural records held in Sanur, it is possible to tell that the style of the buildings, entrance ways and courtyards remains faithful to what might be regarded as a heritage that was once more widespread in Bali. This is in itself interesting because it suggests that Penglipuran, though originally of highland origin, has taken on more mainstream Balinese features. In some other respects, however, the village has remained loyal to its upland origins, even to the extent of being 'ceremonially bare', as Mead put it (1942, p. xiii).

In view of the popularity in Bali of drawing attention to famous foreign residents in marketing tourism, there are scarcely any references to Margaret Mead and Bateson in either Penglipuran or its forebear, Bayung Gede. The anthropologist's sojourn is well remembered and the site of their residence, now a schoolhouse set beside ash-choked streets, is readily pointed out by the villagers, though it has not become a tourist destination, despite Mead's fame in particular. One villager, I Gedar, who worked for the anthropologists as a domestic help, recalls Mead taking numerous photographs, including some of himself, and remembers her for her kindness. This lack of interest in promoting the association of Mead and Bateson with Bayung Gede on behalf of the Balinese authorities is difficult to ascertain precisely, but could have something to do with the setting and the absence of an appropriate dwelling. Artists, such as Walter Spies who lived on the island in the inter war years, are often featured in promotional materials, possibly because their homes are usually attractive and make good tourist attractions. The unwillingness to draw attention to Bayung Gede may also be associated with the village's comparative underdevelopment, and a common assumption in Indonesian officialdom that one should only display the most attractive aspects of one's culture to foreigners. Penglipuran has paved pathways and drains and is shaded by foliage, whereas Bayung Gede has dusty and treeless streets, though this could be improved and would not conceivably discourage tourists interested in anthropological heritage even in its present form. In view of the close linkages between the villages, it would not be unreasonable to propose that both settlements could be marketed and interpreted with reference to Mead and Bateson's research.

The villagers of Penglipuran have been subjected to official interference, but the villagers appear not to be resentful and have responded opportunistically with even a competitive edge to their pride in their 'traditional' status; many do not want to follow the example of other villages since being 'traditional' is a mark of status. Through their retention of control over customary law, the council of elders has helped conserve many intangible traditions, although these are not static and may be modified in accordance with traditional practice. This is not to say that there has been no accommodation with the state since the village does follow national practice in electing a village headman and accepts that there are areas where the state's jurisdiction will apply.

Figure 4.1 The main street and temple in the upland village of Penglipuran

In the absence of much interpretation tourists are unlikely to appreciate why Penglipuran is distinctive from the customary law perspective, and it is the village's tangible heritage, which is accessible and attracts their attention. The fact that the village is 'traditional' is obvious to the Balinese and may well be understood by other Indonesians who are familiar with the concept of 'customary law', but thus far the villagers have not managed to devise a strategy for explaining to overseas visitors how their adherence to customary law makes their settlement 'traditional'. Some interpretation of Mead and Bateson's work could be of benefit here.

There is also a pressing need to update the question of 'aboriginality' into the way the upland Balinese are presented to both domestic and international tourists since it is misleading. The mountain Balinese' designation as 'aboriginal' may well be easily understood by some of the key actors – local elites, guides and guidebook authors – but the theoretical base for this perspective is academically outdated. These cultural differences between lowland and upland Balinese are rather more complex, and are related to variations in the observance of Hinduism that would be challenging, but not impossible, to interpret for tourists. Archaeologists increasingly have to be mindful about the way the sites that they have worked on can become objects of tourists' curiosity, and should not anthropologists also be concerned about making more up to date versions of their theories accessible? The question of ethnographic pseudonyms also concerns issues such as tourism and development, and although there are good reasons for making use of anonymity, there may well come a time when those people want to be known by their own name. Tourism-associated development may well be stimulus for re-designation – or re-branding in the case of locations whose names have not been concealed – but whether or not such people want to be associated with the heritage of a discipline is a matter for them to decide. In view of the pressing need to both attract tourists in the aftermath of the bombings and to address the long-term under-development of Bayung Gede, the work of Mead and Bateson may well have some contemporary applications. Ultimately it is up to the villagers themselves to decide whether the development benefits of being associated with anthropological luminaries are worthwhile.

Given that tourism development in Penglipuran was a top down initiative, involving perhaps even the President himself (Aryantha Soethama 1995, 44), it is worth asking why anthropologists and government are drawn to similar kinds of settlements, and whether some kind of global aesthetic has intruded. It seems unlikely that this was an isolated case in which the work of anthropologists and the interests of government overlapped and if so, can underlying assumptions be detected in what piqued the interest of both parties? In view of the fact that politicians, public servants and anthropologists are often broadly products of similar social systems, notably Western-style education, perhaps we should not be surprised that should be curious about certain kinds of societies and respond to similar aesthetic criteria. In an 'ethnographic confessional' Crick noted the similarities between what anthropologists do and what tourists do and found it a 'painful experience' (1985, 76); perhaps Crick's observations should also be extended to include anthropologists and government.

The Bittersweet Experience of Tanjung Benoa

Another village that has been drawn into tourism's orbit in its own distinctive way is Tanjung Benoa, which is located in the southern part of Kuta in the Balinese district of Badung, close to the upmarket resort of Nusa Dua. The settlement is predominantly Hindu like the majority of the population of the island of Bali, though there are Buddhist (predominantly Chinese) and Muslim (mainly of Makassar descent) communities. With its mosques, Hindu and Buddhist temples, as well as seaside location, the village itself is increasingly being featured in tourism promotional literature. It is, however, overshadowed by its neighbour, the internationally renowned resort of Nusa Dua, which has long been heavily promoted. With its wide sandy beaches and uninterrupted views over the Indian Ocean, Nusa Dua has attracted international hotel chains as well as developers based in the Indonesian capital of Jakarta.

The development of tourism in Nusa Dua, particularly in the latter part of the Suharto era in the 1990s, has not been without controversy, not least because of tensions between local residents and the major hotel owners. There were various fault lines, notably the question of access to the beach for Balinese funeral processions and the construction of a wall to screen off the local residents from the resort. According to the local authorities, these tensions have been resolved and the Nusa Dua Development Authority is upbeat about the contribution made by tourism to the local economy and its perspective is made widely available in published documents. The statistics provided by these outputs, though doubtless as accurate as can be given the context of information gathering within a developing country, do not really provide much insight into the economic impact of tourism within the community living closest to the resort. The purpose of this study therefore was to shed light on how tourism benefits the local residents and in particular how it impacts upon the poorest members of this society, something that cannot be deduced from the macro level statistics provided by the development authority. The aim was to see how local residents perceived tourism and to try to elucidate their views on its merits on helping the poor. The researchers were able to contrast these views with those of the resort since they had participated in a seminar in Nusa Dua and the opportunity to hear how the Nusa Dua Development Authority saw its role and had been able to question one of its representatives before the study took place.

By Indonesian standards Bali has been a haven of relative prosperity, thanks in part to the role played by tourism, but it remains a society in which poverty has not been eradicated. Balinese people share a deep sense of responsibility for how they manage their circumstances, particularly with how effectively they provide for their families. The Balinese also have to fulfil certain financial obligations with regard to their religious beliefs and an inability to pay the tariffs demanded by one's clan for temple maintenance can lead to a deep sense of shame. The local papers regularly report on the suicides of islanders unable to fulfil their social and financial obligations. Poverty remains a pressing concern on the island but is rarely addressed with regard to the way tourism is managed.

In Tanjung Benoa the head of the *banja*r opened the meeting and explained in Indonesian its purpose and how it would be conducted. He raised the re-worked questions that had put been submitted in the briefing notes. One of the Udayana University researchers then went over the research questions in detail. By and large the participants stuck to Indonesian, though they occasionally lapsed into Balinese, the local language, which was fortunate since the two foreign researchers were both competent in the national language. A lively discussion ensued with each member of the *banjar* speaking in turn, though a few declined to comment. Hand written notes were taken by three of the researchers and these were later collated.

As might be expected there were differences of opinion on the role of tourism in their lives, but there was broad agreement around a number of issues, which were largely regarded as positive. First, they were aware of the diversity of economic opportunities provided by tourism, not only through direct employment in the resort in the hotels, but also in a range of related occupations outside or close to the resort such as retail (clothing and souvenir shops), hospitality (locally owned cafés and hotels) and land and street maintenance. Second, there was tourism's general knock-on effect in the economy that improved the quality of life through increased educational opportunities and improved living conditions. An important consideration for the predominantly Hindu villagers was the enhanced ability of people in gainful employment to contribute to the maintenance and functioning of the temples. One participant was of the opinion that small locally-owned companies were the best source of revenue for the village, but he qualified this by saying that the experience gained in the resort, such as learning languages proved useful in running an art shop. In a similar vein, another participant stressed the role of education in enhancing access to the benefits of tourism, presumably through being able to compete for better-paid jobs and providing higher value-added services. With a reference to the Bali bombings of 2002, which was still fresh in people's memories, another cautioned against being too optimistic about the importance of education in helping people to derive benefit from tourism: 'No amount of professors can help solve anything if Bali is not safe!'

While virtually all those who spoke recognised the contribution made by tourism to their livelihoods, what were seen as the negative aspects of living close to a major resort generated more impassioned discussion. The problem of access to the beach had not entirely vanished and there were calls for a more equitable distribution between the village and the resort. There were also worries about the burdens placed on the villagers for cleaning and maintaining the beach, and the consensus was that the resort should be taking a bigger responsibility. One participant complained that, although the locally built canoes were affordable and could enhance one's economic activities through fishing and providing transport for tourists, there was little room for them to be beached. Another participant enthusiastically supported the complaints made on behalf of boat owners, but pointed out that similar restrictions curtailed the economic activities of micro businesses, in this case taxis. He pointed out that as a taxi driver living in a village close to a major international resort, it seemed strange that he had to go all over Bali in search of work. This was because the hotels simply contracted the work to taxi companies on a competitive tender basis with little regard to the needs of small operators living on their doorstep. For at least one participant, it was lack of access to tourism for local businesses that prevented income from international tourism from entering the village in sufficient qualities. Another complained that bureaucracy also impinged upon the benefits derived from tourism and that this study provides a rare opportunity to air grievances about tourism. Most public consultation exercises, he opined, were ineffectual: 'You can say as much as you like, but it will have little impact.'

Education was a recurring theme throughout the discussion with one participant complaining that the multi-lingual talents of the village's schoolchildren not being recognised and built upon. Lack of educational opportunities heightened inequalities and led to a perception that tourism was often unfair in the way that it was managed. The debate about education also concerned conservation with one speaker contrasting Tanjung Benoa unfavourably with the upland village of Penglipuran. The latter village was transformed into a conservation area by an edict of the provincial governor, which laid down strict rules protecting the architectural integrity of the village. In contrast, a variety of building styles had been allowed to flourish in Tanjung Benoa, which, according to this participant, undermined the integrity and attractiveness, presumably for tourists, of the village.

The above discussion was mainly concerned with the role of tourism in the local economy of a Balinese village and was not explicitly concerned with poverty. Poverty cannot be openly discussed in contemporary Balinese society, but concerns about inequality underpinned much of what was said in the focus group. Much of this can be characterised as a lack of access and opportunity, particularly with regard to education, but there are some important caveats with regard to how this is perceived in Bali. For the Balinese, fulfilling one's religious obligations is an important indicator of one's quality of life and being unable to do this brings about an acute sense of anxiety. The Balinese also have to observe a complex cycle of religious festivals and holidays, which may render them temporarily unable to work, but how this can be squared with the demands of an industry with totally different concepts of time and rewards remains to be addressed. It remains unclear how well villagers like those of Tanjung Benoa understand the industry that plays such an important role in their lives since the local people only appear to be involved in the lower and middle strata of economic opportunities and seem to be excluded from the upper levels where decisions are taken. This exclusion could in part be derived from the difficulty of being involved in key roles in tourism while fulfilling important social and religious obligations. Such pressures are felt keenly by the least privileged, and it is clear from the results of this study that discussions to help to address these problems are long overdue.

Debates about beach cleaning and space for mooring canoes provide an opportunity for those who are least able to articulate their positions to indicate their dissatisfaction, but one wonders whether or not this captures satisfactorily what really afflicts the least privileged members of Balinese society. This leaves a number of questions unanswered, which may be summarised as follows: first, are indigenous Balinese unable to realise their full potential in the tourism industry because their religious obligations keep them away from work in ways that are difficult to predict and accommodate? Second, do the demands to fulfil religious obligations fall more heavily on the poor and how are these financial obligations managed? Third, how can the application of international market forces in resorts such as Nusa Dua be squared with a deep sense of equality and fairness that pervades Balinese society? Fourth, what practical steps should be taken to ensure that the Balinese are educationally equipped to deal with an industry that plays such an important role in their lives?

Desa Peed: On the Fringes of Tourism

When the island-wide purification ceremonies were held in the aftermath of the 2002 bombings, sacred water was offered in libations from the holiest shrines in Bali, and one of these sources was the village of Peed on the island of Nusa Penida, which lies 14 km to the east of Sanur Beach. Given that the outrage involved tourists it is somewhat ironic that Peed should be part of the salvation since the island of Penida is only peripherally involved in tourism despite its proximity to and clear visibility from one of Bali's oldest resorts. In developmental terms Nusa Penida is the poorest part of the regency of Klungkung and is struggling with the pressures of deforestation and erosion, and out migration among the young. As compared with the main island, Nusa Penida is noticeably arid, more reminiscent of the islands further east in Nusa Tenggara than Bali itself. Penida has some water and it is reasonably clean, but is difficult to access and expensive to pump. The population of the island is less than 60,000 and is falling.

Nusa Penida is a craggy limestone island, where the inhabitants survive by cultivating cassava, teak trees, papaya, maize, jackfruit and other fruits, with the sea providing fish. The islanders keep the distinctive Bali cattle, which forage for grass and leaves in the barren landscape, it being too dry for water buffalo that are popular on the mainland. The fields are rain fed and not irrigated like elsewhere in Bali and the terraces are made of stone not mud and are reputed to be ancient. Cattle are used to plough the fields but one of the main tools of cultivation is the hoe. The cattle have the advantage of being disease free and, following an epidemic, a Malaysian businessman attempted to buy all the island's cattle, but he was turned down. A mainstay of the island's economy is the cultivation of seaweed, which is grown around stakes driven in to the seabed with lines pulled taught between them. The islanders gather up the weed, half stooping in the water, which must be backbreaking, and it is dried in the sun along the roadsides, whence it is taken to Korea and Taiwan to be turned into cosmetics and fertilisers. Houses with simple palm leaf walls line the shoreline, and serve as temporary dwellings for the seaweed gatherers.

The islanders share some similarities with the Bali Aga (Hobart et al. 1996, 66), though there are members of the titled castes living there. The island's most important temple, *Pura Dalem Penataran Ped*, is located in the village of Ped and is dedicated to the legendary demon, Jero Gde. Worshipers from the mainland come over from the mainland to mark the temple's *odalan*, anniversary festivals, visiting a series shrines linked together by a path, calling at each in turn. There is also a temple devoted to Baruna – one of two on the island – that is associated with the island's maritime activities. On the island's highest point, Puncak Mundi, some 600 metres above sea level, there is a shrine that is said to have originated in the 12th century, which lies on a line of spiritual strength flowing through the temple at Peed to Bali's mother temple, Besakih, and Bali's highest peak, Gunung Agung.

Figure 4.2 Floating restaurant close to the offshore island of Nusa Penida

Invoking the spirit of Jero Gde, who was said to have shown signs of restlessness in the months leading up to the 2002 explosions, the islanders express resentment about their lack of development. They feel excluded and have the impression that the mainland authorities look down on them for being stupid, lazy and lacking in creativity, and that they are only suitable to be servants. The lack of appropriate harbour facilities keeps all but the most intrepid tourists away, and, though the authorities built a new port it was botched and remains unused. The islanders are keenly aware of the importance of tourists and can actually see them from Nusa Penida since companies such as Quicksilver bring them to a floating restaurant close to the island. The food and drink for these facilities is, however, brought from the mainland and few, if any, of the islanders are employed in this venture.

The island's intelligentsia, its schoolteachers and government officials, advance an interesting argument in favour of tourism saying that it would help in the island's conservation efforts through reducing reliance on tourism. The Dutch appear to have instilled an awareness of the perils of intensive farming through encouraging the construction of temples to preserve woodland. Temples are usually surrounded by foliage, which local farmers are unwilling to cut down out of respect for the shrine. The Dutch also inaugurated some practical measures, including building a concrete lined reservoir.

Local officials also cite the example of the neighbouring island of Lembongan, which has a more developed tourism industry, and shows signs of the regeneration of its forest cover. Lembongan is also noteworthy in another respect since some of its inhabitants have foreign spouses, which may be partly due to its accessibility as compared to Nusa Penida. One of the main reasons for Lembongan's relative greenness seems to be the switch to reliance on kerosene for cooking instead of firewood, a commodity that remains relatively expensive on Nusa Penida due to the cost of importing it. Instead of worries about the impact of tourism, one encounters a distinct enthusiasm for what are seen as the potential benefits of this industry, and impatience with the mainland's authorities for not helping it to develop. Such views have a distinctively global feel about them, although the concerns behind them are decidedly local.

Chapter 5

Street Traders and Entrepreneurs

References to Bali as the 'Aussie Costa' crop up in the travel sections of both European and Australian newspapers and magazines, drawing a parallel between the intensity of mass tourism in Spain and the huge annual influxes of sun drenched Australians to this Indonesian island. Like Spain, Bali has sucked in not just seasonal visitors from wealthier nations, but also longer-term migrants from such countries, including businessmen, models, sports personalities, pop stars and artists, as well as second and retirement home seekers. Albeit on a much smaller scale, Bali has got its equivalent of the 'people of the Pueblos' who inhabit the stuccoed rows of cottages that hug the Spanish shoreline, though the Balinese variety has yet to acquire such a damning moniker. At first glance the term 'Aussie Costa' would appear to be appropriate, but in reality the situation is far more complex with not only Australians visiting in large numbers, but also Japanese, Koreans, as well as people from the European Union. Look more closely at the international visitor arrival statistics and a more diverse picture emerges with a huge array of countries of origin being listed, ranging from older sources of tourists such as the USA and newer ones like Brazil and China. Such is the scale of visitation that Bali is well served with honorary consulates catering not only to the tourists but also to the expatriates, the most renowned of which, on account of its humanitarian efforts during the Bali bombings, is arguably the British consulate close to the *Cat and Fiddle* folk club in Sanur.

Impressive though the diversity of international visitors and residents may be, this is only a part of the overall picture, since what is striking about Bali is the way that it has acted as a magnet for arrivals both in terms of tourists and longer term migrants from elsewhere in the huge and ethnically complex nation of Indonesia. South Bali in particular has become a veritable Indonesia in miniature, and though the number the number of business people from around in the archipelago has declined, especially the petty street traders, since the bombings of 2002, the island remains ethnically diverse in indigenous terms. While a great deal of tourism development in Bali has been supported by global capital that has facilitated the growth of enclaves, what should not be overlooked is the importance of Indonesian conglomerates based in such burgeoning cities as Jakarta and Surabaya, and a wide range of domestic entrepreneurs both from Bali and elsewhere in the archipelago. Indonesian entrepreneurs are involved in a variety of tourism concerns ranging from the Jakarta conglomerates to medium-sized firms and small-scale operations, and moreover a clear line cannot invariably be drawn between the Jakarta conglomerates and transnational capital.

Ethnicity permeates many aspects of tourism and related businesses in Bali, but many of the ethnic groups involved have not received a great deal of attention in the literature. An exception are the in the Chinese who are the subject of an increasing number of research papers, but this coverage tends perhaps to overshadow the entrepreneurial activities of the others. Of course any discussion of business in Indonesia would be inappropriate without due consideration of the Chinese, but what is not often investigated fully is their relationship with other groups, something that is a feature of their business activities in Bali. The Chinese have long had an interest in Bali's tourism industry and in fact the Sari Club, which was bombed in 2002, was Chinese owned. The Chinese are famous for doing business with compatriots, which helps to reduce risk, especially when supply lines are extended, but they are prepared to take risks by having dealings with non-Chinese. There is a substantial informal sector in Bali, often financed by Chinese capital, but operated by non-Chinese indigenous (*pribumi*) labour.

Tourism, Entrepreneurs and Development

There is a substantial literature on the role of tourism in stimulating growth in the Lesser Developed Countries. Early commentators tended to draw attention to the potential of tourism to stimulate a variety of economic activities, and this was later superseded by a more cautionary stance. International tourism in the developing world was limited in the 1950s and 1960s when the development model known as 'modernization theory' became widely adopted, and tourism was easily accommodated into the modernization perspective which regarded culture as a barrier to development, and contact with the so-called 'modern' world a solution (Wood 1993, 50). The attack on modernization theory by dependency theorists and others in the late 1960s and 1970s drew attention to the problems of over reliance of tourism on external capital (de Kadt 1979; Wood 1993, 52–4).

The major tourist flows and associated capital emanate from the developed economies, often creating enclaves of commercial activity that exist apart from the main economy of a Lesser Developed Country. This perspective may be likened to situation envisaged by Boeke, in which a sharp cleavage divides society into dual segments: one Western, modern and capitalist, the other traditional and non-capitalist (Boeke 1953). Such tourism zones, so the argument goes, are managed by global capital and transnational organisations, and, unless subject to strict regulation by the local state, permit only limited economic benefits to accrue to the host destination (Shaw and Shaw 1999, 68). In these contexts it may be appropriate to talk in terms of powerful cores and dependent peripheries, bearing in mind that neither are immutably fixed in an historical or geographical sense (Selwyn 1996, 6). Competitive pressures have been accentuated by globalizing trends within tourism, particularly on domestic hoteliers, since the world's best corporations can enter almost any market at any time (Go and Pine 1995, 11). Cleverdon forecasted, however, that not only would the bigger companies grow, but that the small specialists would flourish, though the non-specialist and mid-scale operators would suffer (Cleverdon 1993, 84).

International capital pervades many aspects of tourism, but the emphasis on metropolitan control via transnational corporations in the literature risks overshadowing other kinds of entrepreneurial activity. Global tourism corporations have penetrated many developing countries with foreign ownership accounting for 60 per cent of hotel beds in countries such as Kenya (Rosemary, 1987), but the picture is not universally pessimistic. Kadir Din, for example, cites the case of five internationally competitive hotel groups in Malaysia, which began as local, family enterprises (Kadir 1992). A more recent study in the Indonesian island of Bali has suggested that the picture is quite complex with differences in entrepreneurial activity being partly linked to geographical and managerial features (Shaw and Shaw 1999, 80). Open resorts often provide local traders with opportunities – either formal or informal – to secure a share of the market, whereas more tightly controlled enclaves restrict, often unsuccessfully, interaction between tourists and indigenous entrepreneurs. Generally speaking enclave style tourism facilities marginalise local entrepreneurs both geographically and economically and this in turn increases the size of the informal sector (ibid.). Shaw and Shaw's research suggests that parallels may be drawn elsewhere, notably the West Indies, with regard to the participation of local entrepreneurs in enclave resorts, though the comparison should not be taken too far because the cultures and contexts differ.

The distinction between local and transnational entrepreneurs, for example, is not easily maintained in Bali since there are considerable links between international capital and conglomerates based in Jakarta, the Indonesian capital (Aditjondro 1995; Picard 1996). Indonesia is moreover ethnically highly diverse and what may be assumed to be a local entrepreneur in a given tourism destination may in fact be an outsider, though still an Indonesian citizen, drawn from the furthest reaches of the archipelago. The skills associated with tourism are also not evenly spread and some ethnic groups participate in the economic activities associated with tourism more than others. In Indonesia ethnicity cuts across issues such as local participation in tourism, the role of small and medium-sized enterprises, tourism entrepreneurship, labour relations and labour mobility, but remains largely under researched. This chapter takes Shaw and Shaw's (1999) analysis of local entrepreneurship in Bali as a starting point, but focuses more closely on questions of ethnicity on the one hand, while considering these issues within the broader context of Indonesia and elsewhere on the other.

The term 'entrepreneur' is used here to refer to a person who undertakes a commercial venture, often involving risk. Entrepreneurship may be regarded as a profit-seeking directed towards resolving ill-defined problems in complex and structurally uncertain contexts. Entrepreneurs, however, are not driven solely by profit since the desire for prestige and the constraints and obligations of membership of a particular group (for example a kinship group) may also influence behaviour (Dahles 1999, 13). Entrepreneurship is a central concept in economic theory, but has begun to attract interest in anthropological and sociological circles, and the definition has expanded to encompass other concerns such as management style and personal attitude (Wall 1999). Attention has been paid to the characteristics (for example competence, leadership) of entrepreneurs, organisational behaviour and the role of markets, but the cultural contexts in which these phenomena occur have not always been systematically addressed.

A comprehensive account of all the ethnic groups involved in entrepreneurial activities linked to tourism in Java and Bali lies beyond the scope of this chapter and thus the coverage is limited to the leading participants. The Overseas Chinese are the subject of an increasing number of papers devoted to business and management, and this coverage tends to overshadow the entrepreneurial activities of other ethnic groups. The Chinese, therefore, are not singled out in this paper for special attention and their business networks are discussed alongside those of other ethnic groups. The recent series of economic crisis has led to job losses and expatriates on dollar-based salaries are being replaced with local labour. These changes reinforce the need for studies on the role of ethnicity in the Indonesian tourism economy. Anthropological and sociological perspectives show that ethnicity is flexible and negotiated, and this chapter draws attention to the dynamic character of ethnicity in entrepreneurship.

Plural, Primordial and Situational Ethnicity

The situation in Bali is highly varied with diverse ethnic groups participating in a myriad of occupations associated with tourism. Anthropological and sociological theories of ethnicity enable us to examine the maintenance and extension of ethnic boundaries, and the processes by which networks include outsiders. Tourism brings into contact people who are not only strangers to one another, but are also members of different ethnic groups. The use of the term 'ethnicity' varies greatly, and it is the concern here to examine the theories developed by social scientists and to show how they may be applied to the study of entrepreneurship in tourism. A useful starting point for this discussion is Furnivall's work on the ethnic division of labour in Asia and his concept of pluralism.

Furnivall defined a 'plural society' as comprising two or more components or 'social orders' who lived alongside one another without combining into a single political unit. The different peoples in the colonial societies described by Furnivall were brought together in the market place under the control of state systems dominated by one of these groups (Barth 1969, 16). This may be likened to the situation described by Smith in the Caribbean where there was not a single society, but several societies alongside one another, each with its complete set of institutions (1965, 80). Each institutional set was, however, incomplete because none of the distinct societies had its own political institution and therefore responsibility for order devolved to the dominant group. Based on his experiences in Burma and Java, the former colonial administrator, J.S. Furnivall concluded that the various groups that he observed were not bound together by normative bonds.

> In Burma, as in Java, probably the first thing that strikes the visitor is the medley of peoples – European, Chinese, Indian and native. It is in the strictest sense a medley for they mix but do not combine. Each group holds to its own religion, its own culture, its own ideas and ways. As individuals they meet, but only in the market place, in buying and selling (Furnivall 1968, 304).

There is another group of theories, which may distinguished as the 'primordial' and 'situational' or 'instrumental' approaches (Rex 1986, 27) The first of these perspectives, the 'primordial' view, sees ethnicity as dependent on a series of 'givens': by being born into a particular community, by adopting its values (for example religion) and speaking its specific language, or even dialect of a language, and following a set of cultural practices that are associated with that community

(Geertz 1963,109). We are bound to our kinsmen, neighbours and fellow believers not only by personal attraction, common interests, moral obligations or tactical necessity, but also by the importance attached to the tie itself. Social mobility does not change this sense of ethnicity, as can, for example, happen with social class '... since society insists on its inalienable ascription from cradle to grave' (Gordon 1978, 71). Ties of ethnicity are not closely associated with class or political phenomena, and usually crosscut them. These links have their own vitality and internal dynamism and thus exist independently of other elements in the political process (Rex 1986, 27).

In contrast, the situational or instrumental perspective offers a more dynamic view that places emphasis on ethnicity as a set of processes and social relations, which may be invoked according to circumstances. Barth is generally regarded as the proponent of this view and what distinguishes him from earlier theorists is his rebuttal of the notion that ethnicity is a stable entity. Barth raises the important question of where the boundaries between different groups might lie (Guibernau and Rex 1997, 7) and refuses to see ethnicity as the property of cultural groups. By concentrating on social relations between and within ethnic groups Barth, and later Eidheim (1971) and others developed a set of concepts for analysing interpersonal ethnicity (Eriksen 1991, 128).

The situational approach places emphasis on ethnicity as a set of social relationships and processes by which cultural differences are communicated and maintained. In order that an identity may be understood, it has to be constantly invoked through intentional agency. The signifiers of identity may vary in different contexts, but, despite its flexibility, ethnicity appears to be more stable than many other markers of individual and group identity. If expressions of ethnicity are to be understood in interpersonal encounters, then there must be shared agreement concerning what is significant. The social communication of cultural difference may be observed and described, though these activities are elusive and difficult to quantify analytically, not least because ethnicity cannot be reduced to a fixed system of signs (Eriksen 1991, 130). The pursuit of cultural identity and social autonomy does, however, involve the manipulation of symbols as boundaries are defined and maintained; ethnicity may be seen as a resource at the disposal of entrepreneurs as they structure and restructure their worlds, mobilize resources and build networks in a competitive environment.

Balinese Entrepreneurs

In view of their long exposure to tourism, it is not surprising that the Balinese are involved in a wide range of entrepreneurial activities associated with this industry. The Balinese, often from a low capital base, have started many small and medium sized concerns. Many Balinese entrepreneurs are women who developed successful concerns, particularly in hospitality, often with little formal training (Mabbett 1987). Made's Warung in Kuta and Murni's Warung in Ubud are both examples of businesses run by women that were started with a modest capital outlay. They devised menus on a trial and error basis without risking too much of their initial investment. Competitors, however, often copy successful formulas, forcing the entrepreneurs to carry on innovating. It may be argued that much could be learned from the Kuta experience with regard to its low capital entry costs in the late 1970s (Hampton 1998, 651) since it is widely held that small-scale developments have modest requirements which enhance local participation and are thus associated with higher multipliers and lower leakages. The counter argument, however, holds that large-scale developments are necessary to ensure economies of scale and access to international markets via effective marketing (Wall and Long 1996, 33–34). The possibility that the success of the small and medium-sized businesses in Bali may be partly due to the presence of large-scale tourism developments merits further investigation.

A hotel that was specifically designed with tourists in mind, the Bali Hotel, was opened in 1928 in Denpasar. It was located in South Bali in what was seen as the most culturally vibrant part of the island and, in order to entertain the guests, the management organised weekly performances of Balinese dancing (Picard 1997, 190). The shows became very popular and were adapted to suit the taste and attention span of foreign audiences. A new kind of dance evolved which became detached from its original religious and theatrical context, and thus more suited to commercial exploitation (ibid., 191).

In addition to the provision of entertainment, the Balinese have long been involved in a wide range of tourism-related activities, notably the retailing of arts and crafts, an example being Jero Nuratni who owned not only a well patronised shop in Denpasar in the 1970s and 1980s, but also outlets at airports, including Jakarta. Such was the esteem with which Balinese arts were held that Nuratni's business became the main supplier of state gifts for the presidential palace in Jakarta. Even before the advent of mass tourism, the Balinese had a presence in business in the 1950s and 1960s, notably Balinese figures like Ida Bagus Kompyang who built and operated the Segara Beach Hotel in Sanur, and I Nyoman Oka alias Nang Lecir who set up a tourist agency in partnership with a Chinese taxi operator.

By the 21st century the Balinese had entered the ranks of what are called locally the 'Kuta billionaires', the best known of which include Kadek Wiranatha, the owner of Paddy's Bar, and his brother Gde Wirata. They also own New Paddy's, as well as several hotels, restaurants, in Kuta and Seminyak, one of which is KuDeTa, a well-known haunt of celebrities that has been reviewed as far away as London. After the bombings of 2002, the brothers established an airline called Bali based *Air Paradise International*, which unfortunately had to close in the aftermath of the 2005 bombings.

Historically, Bali's influence stretched well beyond the island's borders and this continues to complicate its relations with its neighbours. Between 1500 and the mid-17th century the Balinese royal house of Gelgel ruled a kingdom that stretched from Blambangan and Pasuruan in East Java to the western peninsula of Sumbawa. All of neighbouring Lombok came under Balinese rule, though Balinese Hindus were a minority, the majority being indigenous Muslims known as Sasak. Migrants from Lombok – both Hindu and Muslim – as well as immigrants from the eastern tip of Java are often encountered within Bali's tourism industry. Given the historical links between their homelands and Bali, these migrants have recourse to a kind of affiliated

Balinese identity, and their numbers are difficult to verify. By the early 1980s, the Balinese were also playing a leading role in tourism in Nusa Tenggara Barat to the east of Bali, often trading on their knowledge of and historical associations with the area. The Chinese are less influential in this region than they are further west, and the Balinese occupy some of the niches in which they prevail (for example medium sized hotels) elsewhere.

The infrastructure that was introduced alongside tourism also facilitated the development of a range of export-orientated industries, notably handicrafts, which helped to turn the island into a major commercial centre in its own right. By the early 21st century, Bali not only produced a wide range of goods for both the tourist and the export markets, but also acted as a distribution centre for commodities made elsewhere. Handicrafts are traded along the hub and spoke distribution systems of market economies and may involve quite different producers and retailers. Products drawn from throughout the vast Indonesian archipelago may, for example, be purchased in Kuta Beach, Bali, often without any information whence they came.

Bali also acts as a kind of training ground for entrepreneurs who eventually open businesses on other islands; small-scale traders often refer to Kuta Beach as the *Universitas Pantai*, the university of the beach. Cukier's study, for example, shows that the majority of vendors in Kuta and Sanur are single men, usually teenagers and young adults, with limited levels of formal education. Despite the fact that only a minority of them have graduated from high school, many are multilingual and commonly speak their own language, the Indonesian national language, and at least one foreign language, usually English (Cukier 1996, 67). Much of the vendor's success is dependant on his or her ability to communicate with potential clients, and thus mastery of a foreign language is a prerequisite (ibid., 68).

The tourism zones of Bali also attract a large number of entrepreneurs from across the archipelago, many of whom become involved in syndicates controlled by Chinese capital. The most notorious of these are the counterfeit watch sellers of Kuta who flash open their display cases at the sides of streets, taking advantage of any hesitation or interest in their wares on the part of the tourists (Shaw and Shaw 1999, 77). A survey by Cukier and Wall showed that 85 per cent of the vendors in Sanur and 73 per cent in Kuta were non-Balinese (1994, 465). These outsiders came overwhelmingly from the island of Java, 68 per cent in the case of Sanur and 73 per cent in Kuta, though the survey did not specify which ethnic groups they belonged to (ibid.).

A later study by Cukier provided additional information on gender and revealed that the majority of male vendors in both Kuta and Sanur were non-Balinese, whereas the majority of female sellers were Balinese (1996, 69). The research also brought to light a gender division of labour with regard to the products being sold: men only sell watches and sunglasses, whereas women sell garments (ibid., 69). Intriguingly, this divide is reminiscent of a much older division of labour in Indonesia that has been well covered in the literature. Certain tasks are nearly always associated with one gender, the most common divide being the production of cloth by women and the working of metal by men (Adams 1973, 277). Many Balinese, particularly men, hold jobs in kiosks and hotels, and work as guides, occupations that have not been penetrated by Indonesians from elsewhere (Cukier 1996, 72). In these contexts local knowledge of customs and festivals may provide the Balinese with a competitive edge.

Figure 5.1 **Former restaurateur, Tini Pidada, turned to batik retailing and designing under the name 'Dayujiwa' in the aftermath of the Bali bombings**

The Javanese

Java is located midway down the Indonesian archipelago and covers an area of approximately 132,246.4 sq kms. Despite being Indonesia's third largest landmass, Java is the most populous Indonesian island and its 100 million inhabitants comprise roughly half of republic's population. Four main languages are spoken on the island, all of which, like Balinese, belong to the Austronesian family. The descendants of the original population of Jakarta, roughly 10 per cent of the city's 9 million people, speak a dialect of Malay. Sundanese is spoken in the middle and southern portions of West Java, whereas in northeast Java Madurese has been preserved by the descendants of migrants from the island of Madura. Javanese is spoken in the rest of the island, though a distinction is made between the dialect of the northwest (Cirebon to Banten) and the 'real' Javanese on the centre and east. In Indonesia the term 'Javanese' is usually applied to speakers of the 'real' Javanese language (Magnis-Suseno 1997, 14). In order to reduce population pressures successive Indonesian regimes have supported a policy of transmigration whereby the government assists citizens from the more crowded islands to settle elsewhere. The majority of Javanese in Bali are either public servants or business people, the island being too crowded to receive transmigrants. Of particular note in the small and medium-sized business sector, often running shops, bars and restaurants, are people of East Javanese origin from cities such as Jember and Malang.

The Chinese may dominate the Indonesian economy, but the Javanese with their numerical superiority and political influence set the context in which business is conducted. It would, however, be misleading to conceive of the role of the Javanese in purely political terms since Java has a long established market economy dating back to at least the 9th century AD (Christie 1993, 11). The Java Sea was a conduit for the Spice Trade and commercial exchanges between India and China, and trade emporiums were a feature of North Coast society. Commerce may be well established, but Javanese culture is not particularly business orientated, there being marked differences between the more outward looking trading communities of the north coast and the more insular peoples of the interior. Javanese society is also highly stratified and this has implications for the way business is pursued, especially with regard to high-ranking individuals. The Javanese themselves recognise two social classes: the *wong cilik* (little people), the mass of agricultural workers and low-income urban dwellers, and the *priyayi*, the officials and intellectuals (Magnis-Suseno 1997, 15). There is also a prestigious aristocratic stratum, the *ndara*, whose outlook is emulated by the officials. Members of the highest echelons of Javanese society may not admire the process of making money and the work it entails, and this doubtless inhibits their entrepreneurial endeavours (Mann 1994, 115). Other factors impinging on entrepreneurship in Javanese society include an unwillingness to ask questions and challenge superiors and a stress on harmony and an attendant avoidance of conflict. Attitudes like these are not easily incorporated into business practices that rely on taking responsibility and using one's initiative.

Despite the existence of cultural factors that may constrain the emergence of entrepreneurs, Javanese may be found in the ranks of the nation's business elites. Some of the leading Javanese families moreover have important interests in tourism, though they may pursue these activities in conjunction with foreign and Chinese partners. It is widely surmised, for example, that Jakarta based conglomerates associated with the family of former president Suharto have huge investments in upmarket tourism in Bali. Suharto and his late wife originated in Central Java and before his resignation in May 1998 the family were widely believed to be the richest *pribumi* family in Indonesia.

The Balinese have not invariably welcomed these investments, and the former Governor, Ida Bagus Oka (1988–98), has been widely criticised for favouring the interests of the Jakarta conglomerates and foreign investors. He was seen as selling off the island to foreign interests, though the Jakarta companies cannot strictly be regarded as overseas investors. The involvement of the Jakarta conglomerates in tourism in Bali may be compared to the situation in Thailand where tourism is found in many peripheral areas, though it is usually the urban entrepreneurs who appear to derive the greatest economic benefit (Parnwell 1993, 239). The scale of the Suharto family's business dealings have understandably attracted a great deal of publicity, but it would be misleading to conceive of Javanese entrepreneurship solely in terms of the Jakarta conglomerates. There is an emerging literature documenting the involvement of the Javanese in a wide range of businesses associated with tourism, though this is difficult to quantify with any certainty. Hampton, for example, has studied the development of small-scale hotels in Yogyakarta that cater for the backpacker market (Hampton 1998b, 12–16).

By the late 1990s recruitment to the ranks of street traders was increasingly along ethnic lines, many of whom were of Javanese origin, notably immigrants from the little island of Raas lying between Madura and Kangean. The Raas islanders succeeded in dominating the street and beach trade in hats and counterfeit watches, in contrast to other Javanese selling ornaments, bracelets and necklaces, and the Sasak from Lombok selling kites, toys and watches (De Jonge 2000, 79). Officially this informal trade was banned, but was usually tolerated, but every now and again the *Hansip* (home guard) or police rounded up illegal vendors. In the best scenario the trader would have his merchandise confiscated, but in the worst he would be expelled from the area, though these measures were largely ineffective since the vendors continued to return. In former times the Raas islanders, who have a long tradition of migration, would travel to Kalimantan, Sulawesi and other parts of eastern Indonesia, but by the late 1980s Bali had become their preferred destination.

Described by de Jonge as opportunists who were prepared to take on anything that comes along, the islanders held courage, physical strength and friendship in high esteem (idid., 80). Migrants from Raas were prepared to take on risks and, unlike vendors from other ethnic groups, were not deterred by periodical raids. They were also adventurous, clever negotiators and relentless in pursuit of customers, biding their time and playing all the tricks of the trade to make a sale. The Chinese businessmen who supplied their wares valued their cunning as traders, and preferred them to other ethnic groups on account of their high sales volume, reliability and solvency (ibid., 81). Generally, they honoured their financial obligations and rarely ran away, and there was strong cohesion within the group, as they all originated from the same place and often knew one another. These vendors mostly lived in a dormitory town, out of the sight of tourists, where thousands of migrants from elsewhere in Indonesia had settled, though the Raas preferred to live in Madurese neighbourhoods and were not keen to live among other ethnic groups. They lived in cramped conditions among relatives in a rented or purchased house with singles renting a room in a guesthouse or dormitory and, though most dwellings had electricity, many were dependent on public wells and taps for water (ibid., 83). Many Raas returned regularly to their island, saving money to spend on luxury items rather than improving their living conditions in Bali, and almost all of them would go home for the Islamic fasting month to celebrate Idul Fitri, the feast at the end, among relatives and neighbours. The Raas islanders occupied the lowest rung of migrant society, often seen by other Indonesians as stereotypical Madurese: ill mannered, impudent, untidy, religiously fanatical and even dangerous. Other ethnic groups may have kept apart from one another, but the distance between the Raas and other Indonesians appeared to be considerably greater.

The Indonesian Chinese

The Chinese are well known in the region for their skills in finance and networking and they are prominent in the big cities such as Jakarta, Surabaya, and Medan. The Chinese run many of the four and five star hotels in Southeast Asian cities, as well as numerous middle range hotels, and numerous smaller ones in the ubiquitous China towns. Chinese entrepreneurs are also involved in the informal sector, notably as the providers of capital for certain street hawking rings selling souvenirs and other services in South Bali. As mentioned above the Sari nightclub was under Chinese management, thought the site was rented from a Balinese from Legian and the owner of one of the biggest garment industries in Bali, Mama Leon, is also Chinese.

The Chinese are internally heterogeneous and may be subdivided on the basis of dialect. The Malaysian, Singaporean and Indonesian Chinese are predominantly descendants of 19th and 20th century migrants from South China: Cantonese, Teochew, Hakka and Hokkein. An important distinction is also made in the Malay-Indonesian world between the Chinese who speak Malay, Indonesian or Javanese as a first language and are not closely affiliated with Chinese dialect groups. The latter are variously called 'Straits', *Peranakan* and *Baba* (*Nyonya*) Chinese, often the descendants of pre-19th century migrants who have adopted local languages and customs, married into local families, but stopped short of adopting Islam, the religion of the majority of the indigenous population. There is also a small minority of Chinese Muslims.

Indonesia is home to the world's largest population (7.2 million) of Overseas Chinese, though estimates vary on the precise size of the population. Chinese Indonesians have an ambiguous status that goes back to the Dutch colonial period. On independence in 1950 the Indonesians inherited an economy that had been under Dutch control, with the Chinese occupying a subsidiary role as shopkeepers and traders, and acting as middlemen between foreign exporters and indigenous farmers (May 1978, 71). The Chinese rapidly consolidated their position in the new republic and by 1974, the daily newspaper *Nusantara* was able to report that 90 per cent of domestic capital was in Chinese hands (May 1978, 392).

The Chinese are thought to control roughly three quarters on the 140 big conglomerates that dominate the Indonesian private sector (*The Economist*, July 26, 1997, 11). Businessmen critical of Indonesia's regulatory environment in the Suharto period argued that the President's family operated in partnership with Chinese companies in squeezing out *pribumi* (indigenous) businesses (ibid.). It is widely acknowledged that the success of the Chinese is bitterly resented by the indigenous, *pribumi*, population and that the Chinese are subjected to bureaucratic harassment and are obliged to adopt Indonesian sounding names, as well as being expected to carry identity cards to prove their right of abode. The riots that accompanied the downfall of President Suharto are reminiscent of the attacks on Chinese property that occurred towards the end of the Sukarno period in 1965 (May 1978, 135).

A critical problem in post Suharto Indonesia is the integration of the Chinese. The capital and talent of the Chinese need to be harnessed if Indonesia is to prosper, but continuing discrimination undermines their confidence. The ethnic and religious intolerance that accompanied the demise of the New Order regime remains a barrier to reform. Some analysts have argued the Indonesian crisis will be prolonged if economic hardship makes the Chinese minority an easy scapegoat (Godley 1999, 53). The disproportionate economic power enjoyed by sections of the Chinese community under Suharto is an understandable source of discontent, but the violence directed against the community as a whole during the May 1998 riots has led to international condemnation. The attacks have been widely blamed on clandestine military operations manipulated by General Prabowo, but prominent Muslim intellectuals have also suggested that some kind of 'affirmative action' may be needed to redress the economic imbalance between the Chinese and *pribumi* populations.

Bali, with its good air links to the major Indonesian cities, provided many Chinese with sanctuary at the height of the riots. In comparison with neighbouring Java, the island remained largely strife free and tourism and other export industries (for example handicrafts) kept the economy afloat. The largest Chinese contingent appears to have come from Surabaya, though Chinese from other major centres such as Surakarta, Jakarta and Yogyakarta were also well represented. Estimates regarding the number of refugees from the anti-Chinese riots varied enormously from 10,000 to 100,000, but by July 1999 approximately 10% of the population of Denpasar, a city of 400,000 inhabitants, was thought by academics at Udayana University to be Chinese. In the disturbances an estimated US$20 billion may have been moved abroad by Indonesia's Chinese business elites (*Sydney Morning Herald*, 3 December 1999).

Initially many Chinese businessmen simply wanted to find a safe haven for their families while they continued to manage their interests in Java by making use of Bali's efficient communications. But with their property rights still vulnerable in Javanese cities, many Chinese have decided to remain, perhaps indefinitely, on the island. Some have been attracted to the light industrial centre of Gianyar, but many have chosen to remain in South Bali and invest in the industries associated with tourism. The influx of Chinese settlers and their attendant wealth into to an already densely populated and highly urbanised part of Bali seemed likely to put additional strains on the environment and infrastructure. During the anti Chinese riots in the 1990s, the price of land and house rentals soared in Bali due to increased demand, and an interesting by product of this influx was increased demand for an Indonesia-Balinese dictionary, possibly indicating that the Chinese wanted to be more conversant with the local language. The Chinese population may have increased but there is still no China Town, unlike many parts of the world where the Chinese have migrated, suggesting that the Chinese are content to live among the indigenous Balinese.

The terms network and networking are helpful labels for analysing complex business relationships, decisions and strategies that are difficult to verify empirically (Menkoff and Labig 1996, 128–9). The lack of empirical information tends to fuel speculation about the alleged success of this minority, which may perpetuate stereotypes about the social exclusiveness of the Chinese (ibid.). The Overseas Chinese have long appreciated the advantages of trading networks in order to minimise risks, transaction costs and the uncertainties of external economic relationships (ibid., 130). In common with large-scale enterprises, small firms penetrating foreign markets experience particular problems concerning cultural differences, lack of information and unfamiliar legal arrangements. It is vital, therefore, that the overseas partners, sales representatives and agents of these firms are reliable. Interpersonal links in a network enhance the flow of dependable information; may facilitate the use of sanctions to reinforce desired behaviour; and increase the possibility that disputes among members will be resolved fairly.

The 'trusted networks' of the Chinese, are based on two or more persons sharing a commonality of identification in what are known as *guangxi* relationships. Commentators often assume that kinship and family ties form the basis of Chinese *guangxi* relationships because of the moral obligation that relatives should trust and assist one another. In reality, however, non-relatives and even foreigners can be incorporated into one's identification system. Consanguineal and affinal bonds are important in *guangxi* relationships, but so are fictive kinship ties such as god parenting and exchanges of vows of blood brotherhood. *Guangxi* ties can also be established by appealing to the real or imaginary ties between people baring the same Chinese surnames. The possession of common Chinese surnames does not presuppose common ancestry, though early migrants to Southeast Asia often banded together in surname associations to provide mutual assistance (Menkhoff and Labig 1996,134).

Friendship ties developed through co-residence, schooling and other shared social experiences can also form the basis of *guangxi* bonds. The so-called 'locality' ties among Chinese who share a common place of origin, which were particularly important in 19th century Singapore, may also be a source of solidarity. In Singapore, associations based on locality are founded on five administrative tiers: province, prefecture, district, borough and village, and may cut across dialect groups (ibid.). Trading networks based on locality links are found across Southeast Asia, and may facilitate the development of business relationships with those who set up joint ventures or locate new suppliers in China (ibid., 136). Many Chinese entrepreneurs have business interests spanning the region, as is the case with the Malaysian businessman, Robert Kuok, who controls the Shangri-La chain. Likewise, the Bangkok based Sino-Burmese Ho family have invested in hotels in Myanmar. *Guangxi* relationships may also be formed with outsiders who do not share kinship or locality ties with the Chinese. As compared with other kinds of bonds these may be considered to be the least stable, though many Chinese firms engage non-Chinese partners. In a global economy foreign linkages may become an increasingly important consideration for business networks rooted in kinship and locality.

The theory of pluralism is useful when considering many of the *guangxi* relationships utilised by the Chinese, but the fact that non-Chinese can also be incorporated into these arrangements, particularly with regard to the hotel sector, suggests that boundaries are negotiable in certain contexts. The Chinese are capable of mixing in the marketplace while retaining a separate identity, but they are also able to combine with non-Chinese when the need arises.

Other Indonesian Migrants

Some of those involved in business in Bali or in exporting Balinese business skills elsewhere certainly trade on an adopted Balinese identity or connection. These associations may in part account for the popularity of Bali as a romantic destination, something that young Indonesian entrepreneurs are able to exploit. Many of the men (beach boys) and women who engage in holiday romances with overseas tourists are not Balinese, but migrants from Java, Sunda (West Java), Lombok and further a field. It is often only after a liaison has been established that a beach boy or girl's true ethnic origins are revealed. Sometimes those involved in the holiday romance business, whether men or woman, may play down their Muslim faith in favour of a vaguer, and presumably less threatening (to Western tourists), syncretistic identity, often incorporating Hindu elements.

Another important group of Islamic entrepreneurs who have become involved in tourism in Bali are the Minangkabau of West Sumatra and neighbouring areas. Land is inherited matrilineally in Minangkabau society and there is a long tradition of migration, *merantau,* particularly by men, often in search of fame and fortune. Roughly a third of the people identified as Minangkabau live in the outer migrant area known as *rantau* (Persoon 1986). Migrants meet and network in the distinctive Minang restaurants that are found throughout the Indonesian archipelago. Popular with both domestic and international tourists, these restaurants have spread further afield to cities abroad such as Sydney and London. In Bali, especially Sanur and Kuta, there are a growing number of Minang restaurants and their special style of Indonesian smorgasbord, which are open 24 hrs, not only provide people with food but also to keep tourist areas alive at night. The restaurants are easily identifiable on account of their distinctive Minang architecture and the way the food is presented. In Bali, the number of Minang is growing and, as is the case in other parts of the world where they have settled, they have their own community group, Minang Saiyo, which helps to channel business among the members. In a popular Indonesian aphorism 'Padang' is said to be an acronym for *Pandai Berdagang* meaning skilled (*pandai*) in trade *(berdagang*). South Sumatrans belonging to various ethnic groups also work in Bali and their activities range from providing garments for resort boutiques to running sateh bars.

Bali also has a long established Bugis and Makassar community in the north around the port of Singaraja, and in Tanjung Benoa in the south of the island. These descendants of traders and seafarers from South Sulawesi are engaged in businesses associated with tourism, commonly market stalls and guesthouses, but are not as assertive in Bali as they are in eastern Indonesia. Highlanders from South Sulawesi, for example, accuse the Bugis in particular of trying to muscle in on the lucrative tourism business stimulated by Toraja culture, thereby reviving age old rivalries (Adams 1997, 163). Bugis, Makassar and other sub-ethnic groups from Sulawesi are often involved in jewellery and silversmithing, as well as in retailing antique furniture, which is said to be part of the venerable heritage of their island of origin.

Certain ethnic groups have become so closely associated with particular kinds of goods and services that others cash in on their reputation. Many textiles made in the Sumba style, for example, are mass-produced in factories in Java and have no connection with the actual island of Sumba. Sumbanese traders have, however, ventured to Bali to sell textiles, thereby joining the trade circles of entrepreneurs from around the world (Forshee 1998, 111). The development of international craft markets, which depend on the trade and transport facilities provided by tourism, renders the label tourist art inadequate.

Families of Arab descent, particularly in Singapore, Sumatra and Java also run competitive travel and hospitality concerns, and some have opened businesses in Bali and further east. Sharing a common religion with the Malays and Javanese, many Arabs have intermarried with local families. There are prominent businessmen of Hadhrami and Yemeni origin who have family networks stretching as far as Lombok and Bima-Sumbawa. In the Malay-Indonesian world the Arabs are respected on account of their business acumen, and their association with the birthplace of Islam. In the capital of Denpasar, the Arabs mainly live in the main streets around the market where they own shops selling fabrics, and Sulawesi Street is known locally as *Kampung Arab* (Arab Village), despite the fact that there are more Chinese there today than Arabs.

Networks and Ethnicity

In an often uncertain and volatile commercial environment, entrepreneurs seek to reduce their exposure, and thus ethnicity remains a powerful resource in 21st century Indonesia. The unpredictable behaviour of some of the governments in the Southeast Asian region and the poor regulatory climate strengthens the need for trust-based relationships. Many entrepreneurs have widely extended supply lines and thus attempt to minimise risk by dealing with compatriots. In order to cope with uncertainty entrepreneurs often place faith in networks based on ties of kinship and ethnicity. Entrepreneurs also understand the value of ongoing experimentation with regard to the goods and services that they provide, and learn to drop unsuccessful ideas. Networks based on ethnicity may provide useful conduits for the transference of skills and knowledge. Tried and tested ideas may continue to be used for years, and members of the same ethnic origin guard often access to this information. Networks often have boundaries that follow cultural patterns, especially ethnic ones, and thus the ebb and flow of these perimeters may be analysed with the aid of theories not conventionally applied to the economic sphere.

Furnivall's pluralism has analytical value when considering the entrepreneurial activities of the different peoples involved in tourism, but cannot be applied uniformly. The early colonial travellers, for example, who stayed in government residences, often had little interaction with the indigenous people, but the picture is somewhat different when one considers the role of the European small-scale entrepreneurs. The fortunes of Bali's Bohemian hoteliers in the interwar years were after all closely intertwined with those of their neighbouring Balinese, and parallels may be drawn with Western tourism entrepreneurs and their Indonesian spouses in the 21st century.

The relations between the Balinese and Javanese also cannot invariably be interpreted in terms of pluralism. Both ethnic groups may mix as competitors in the marketplace without combining, and there appears to be little overlap between the Jakarta-controlled conglomerates and Balinese enterprises. Yet the Javanese and Balinese are widely regarded as loyal citizens of the Indonesian republic and both groups continue to identify with the common heritage of Majapahit. It was also the former governor of Bali (1988–98) who played an important role in securing investments and favoured the Jakarta conglomerates along with foreign investors.

Successive postcolonial governments in Southeast Asia, have tried to weaken the links between ethnic groups and certain kinds of occupations, but the measures adopted have not often been effective. Indonesia has not adopted such strong affirmative policies in favour of indigenous businessmen as Malaysia has done, but has a history of discriminating against the Chinese. Pluralism theory may be useful when considering the Chinese, but a modified position would appear to more appropriate. The Chinese are capable of mixing in the marketplace while maintaining a separate identity, but are also able to combine with non-Chinese when the need arises.

In a similar vein, the Balinese continue to network along ethnic lines, but are not averse to making strategic alliances with non-Balinese where appropriate. These connections moreover may be drawn into the primordial domain through powerful cohesive forces associated with marriage. What is particularly interesting in the case of the Balinese is the prestige accorded the name 'Bali' itself within the context of tourism and the related cultural sectors. Some non-Balinese entrepreneurs are prepared to invoke a Balinese identity if the situation requires it, though this is largely linked to contexts involving non-Indonesians. With some of these entrepreneurs this would appear to be a relatively easy adjustment in view of their own cultures' historical affinity with the Balinese. Ethnicity may classify individuals according to origins and background at a more general level than other markers of identity such as kinship, but membership of a particular category depends as much on self-identification as identification by others (Barth 1968, 11–13). Occasions arise within tourism where it is advantageous for an individual to claim membership on another group, though it remains unclear how much conflict arises from this uncertainty.

The establishment of networks of cooperation may be regarded as a risk reducing strategy particularly in view Indonesia's uncertain regulatory climate. Within these contexts ethnicity is a powerful asset that can be invoked according to circumstances to develop, inform and expand entrepreneurial activities. Ethnicity also acts as a powerful cohesive force that may help a group to maintain control over vital resources by excluding outsiders. As this chapter illustrates, it is perhaps more helpful to think of the primordial and situational perspectives as part of a continuum rather than mutually exclusive explanatory frameworks. In a similar vein the concept of the plural society has some validity within the context of tourism, though the term 'modified pluralism' might be more appropriate. Above all ethnicity may be seen as a vital resource at the disposal of entrepreneurs as they organize and reorganize their worlds, mobilize resources and expand their contacts in a competitive environment.

Chapter 6

Global – Local Encounters

Like the other Austronesian peoples of maritime Southeast Asia and the Pacific, the Balinese are not unfamiliar with the experience of cross-cultural encounters. Throughout many centuries they have been obliged to create conceptual and practical strategies to deal with the periodic arrival of strangers, sometimes powerful and dangerous, but often simply intriguing (Reuter 2003, 172). These encounters began long before Europeans appeared on their shores wanting to trade and later conquer, and may be traced back to the common historical experiences of these related peoples, namely a history of mobility and of successive waves of migration across the islands of this vast region. The waves themselves are often not historically identifiable, and need not necessarily have involved large numbers of people, but their legacy has been recurring myths of influxes of immigrants encountering settlers who had arrived earlier, and who become the original dwellers, *penduduk asli*, in the eyes of the newcomers. Given that these waves of settlement are usually seaborne then it is often the peoples of the interior, especially the mountainous uplands, who come to be regarded as the descendents of the earliest arrivals.

The Balinese certainly conceive of their identities as a succession of encounters between earlier and later arrivals, and the most memorable of these that remain significant today are often associated with specific historical crisis points. Every group of new arrivals was obliged to acknowledge that others had come before them, but only some of the later migrants were sufficiently powerful and/or culturally astute enough to establish themselves in Balinese society and to wield political power. The mountain Balinese are widely conceived of as the island's indigenous people, descendants of the first settlers, and they symbolically represent an important counterpart to the island's pre-colonial rulers of the coastal principalities. This is because the kings and nobles of Bali identify themselves with exemplary newcomers with an external and identifiable point of origin, the illustrious Javanese kingdom of Majapahit. Whether or not this settlement was occasioned by a huge influx of refugees, the elite of Javanese society – its aristocrats, artists and intellectuals – following the demise of Hindu Majapahit and the rise of Islam, as is popularly held in Bali, remains a mute point. It is the case historically that Bali was absorbed into Majapahit, but this does not necessarily have to have involved large displacements of people since the local Balinese elites could simply have absorbed novel and attractive ideas of Majapahit, coming to believe over time that they were its rightful heirs.

This dualist model of identity – stranger kings, warriors and priests versus upland aboriginals – originates in the idea that some kind of transformative encounter took place concerning the arrival of powerful newcomers on an island already claimed by the indigenous Balinese. Embodied in this scenario was the potential for conflict, military or otherwise, unless the crisis of encounter could have been resolved culturally and intellectually. There is no reason to assume that the first settlers invariably survive the encounter with more powerful newcomers other than as a distant memory; history is full of cases of dispossessed indigenous peoples, marginalised by stronger migrants. But in Bali, as in other Austronesian societies, the people whose forebears first cleared the forests and cultivated the fields are often respected, not least because it is the spirits of their ancestors that continue to protect the land and ensure its fertility. Their descendants thus have privileged access to the ancestral spirits, and while this alone neither ensures their survival nor the maintenance of first settler rights, it does provide options for the integration of the two populations by peaceful means (Reuter 2003, 170).

From the 16th century onwards a very different kind of stranger started appearing on the shores of Bali, namely Portuguese, Spanish, English, French and Dutch sailors, though often little is known about their specific interactions with the islanders. An exception is Cornelius de Houtman's Expedition of 1597, which was the first reasonably well-documented European visit to the island. From the notes of one of his men, Aernoundt Lintgensz, we know that the ship's crew received a very warm welcome indeed from the Balinese, and that Lintgensz and his crew-mates were treated in a friendly manner by the inhabitants of Kuta, which was very different from the reception they had experienced in Java (Agung 1989, 3–4). Another member of the fleet, Jacob Kackerlack, said of Bali that '...the inhabitants treated us in a very good and friendly manner, bringing to our ship water and pigs, which tasted delicious' (Vickers 1994, 5). It was apparent from both accounts that the Portuguese had already visited the island, and that the king himself was prepared to welcome foreigners, although preferably for short visits or to form political allegiances.

Before the onset of colonialism, Balinese rulers and their people continued to welcome foreigners positively, but as the Dutch began to impose their rule on the island from the 19th century onwards, more negative Balinese images of Westerners began to appear. Balinese painters and sculptors started depicting Europeans as *raksasa* or demons, while others showed them as ambiguous figures of fun (Vickers 1984), and by the late 19th century the expression *I Bojog Putih*, white monkey, had become a general and very negative, expression for Westerners, one that seemed to infer they were magically dangerous, since white monkeys are associated with witches in Bali (Wirawan 1995, 102).

Friendship and Mixed Marriages

Visiting artists, musicians and scholars, some of whom became long term residents, provided some Balinese with opportunities to interact with outsiders, and yet again the islanders have had to develop strategies to deal with these potentially risky – but possibly lucrative and interesting – strangers. The encounters that took place against the backdrop of the stirrings of tourism are often conceived as being quite positive by the Balinese intelligentsia, as is illustrated by the stories of the artists, anthropologists and foreign scholars settled in Bali for varying lengths of time, including permanently. Painters such as Walter Spies and Rudolf Bonnet in the 1930s, Antonio Blanco in the 1950s, and Arie Smit in 1960s, were accepted into

Balinese society. Ubud in particular became an internationally renowned centre for the arts as a result of creative interactions between Balinese and Western artists, and the artists from other areas have very fond memories of their foreign friends. This was also the case on the coast at Sanur where the Neuhaus brothers were known collectively as *Tuan Bé*, Mr Fish, on account of their aquarium; Jack Mershon, an American who also lived in the area is also warmly remembered by the local artists (I Ketut Rudin, personal communication 11/12/1996).

Despite the anti-American sentiments in Indonesia during the late 1950s and early 1960s, there is evidence that the foreigners living in Bali continued to enjoy good relations with Balinese at various levels of society. The American anthropologist Clifford Geertz, for example, who conducted fieldwork in 1958, resulting in a famous article on Balinese cockfighting, wrote of his experiences in making contacts with the Balinese. Initially Geertz and his wife Hildred were totally ignored as if they did not exist:

> We were intruders, professional ones, and the villagers dealt with us as Balinese seem always to deal with people not part of their life who yet press themselves upon them: as though we were not there. For them, and to a degree for ourselves, we were nonpersons, spectres, invisible men.
> [...] except for our landlord and the village chief, whose cousin and brother-in-law he was, everyone ignored us in a way only a Balinese can do (Geertz 1973, 412).

Only when they and the other villagers with whom they were staying had to flee a police gambling raid, were they accepted as members of the community:

> The next morning the village was a completely different world for us. Not only were we no longer invisible, we were suddenly the center of all attention, the object of a great outpouring of warmth, interest, and most especially, amusement (Geertz 1973, 416).

The positive features of Balinese-Western interaction described by such writers as Geertz are contributing factors to the development of tourism in Bali. Specific encounters between Balinese and Westerners in fictional works written by Balinese writers in the past three decades, however, present a more complex picture in which conflicts also arise.

Sexual contact and marriage is another important dimension within the interaction between Balinese and Westerners. The existence of sexual liaisons between Balinese and Westerners has been a reality since the colonial period (Lindsey 1997, 29–34), but in order to maintain a distance between the coloniser and colonised, mixed marriage was discouraged (Maier 1993, 44). Western women were prohibited by the Dutch colonial government from having intimate relations with native men, a notorious transgressor having been K'tut Tantri in the 1930s (aka Suarabaya Sue, Muriel Pearson), which she relates in her widely translated autobiography *Revolt in Paradise* ([1956] 1982, 36–38). In contrast, relationships between European men and indigenous women were more likely to be accepted, as was the case with Le Mayeur de Merpres, the Belgium aristocrat and artist, who fulfilled the dream of paradise by marrying the stunningly beautiful dancer, Ni Pollok, in 1935, after she had been his model for three years.

With the rapid development of mass tourism underway, tourism created more space for Balinese and Westerners to meet, love and intermarry. A marriage between an Indonesian or Balinese and a Westerner was usually termed a 'white-coffee couple', *pasangan kopi susu*, because of the contrast of their skin colour. Mixed marriage was one of the noticeable results of changes in the 1970s, though a small number of such marriages had already occurred in the colonial and national revolutionary period. One of the best-known marriages to occur at a time of widening acceptance was that of the dancer Ni Rondji and the Spanish painter Antonio Blanco, who advertised their domestic circumstances with a sign outside their home inviting tourists to meet the artist, his wife and his Balinese family.

By the late 20th century mixed marriages between Balinese and foreigners had become quite common and more socially acceptable than ever before, but there was widespread recognition that not all of these marriages would succeed and that differences in culture and identity, as well as divided loyalties, provided a potential source of friction. In particular, marriages that grew out of holiday romances were seen as potentially fraught, especially when it involved a young man from Bali – or elsewhere in Indonesia – returning to the bride's country of origin only to discover that he had lost the social standing that he had hitherto taken for granted. These husbands often lacked the appropriate technical or educational skills to adapt to their new homeland and ended up being partially dependent on their wives, a humiliating position for these men given the expectation in Bali that men should provide for their families. A popularly held belief is that marriages between foreign men and Balinese women generally fare much better.

Romances Involving Tourists

Given the ubiquity of tourism in Bali, and its intrusion in to many aspects of everyday life, it is not surprising that themes or sub-themes relating to tourists and tourism-related activities have become a powerful source of inspiration for modern Balinese authors. In short stories and novels written during the tourism boom years from the late 1960s to the 1990s, encounters between Balinese and foreigners that took place within the context of tourism are a recurring theme. These works not only explore the impact of Western culture upon Bali's social and cultural fabric but also reveal how the Balinese perceive foreigners and reflect on their own identities through contact with 'the Other'.

When Balinese and foreign characters interact in fiction, it is usually within a context provided by tourism, and in this setting 'foreigners' refers specifically to Westerners: Australians, Europeans and Americans. Asian 'foreigners' are rarely portrayed in creative works published since independence, though they do appear in earlier writings. Panji Tisna's *Sukreni Gadis Bali* (1936), for example, depicts an Indian journalist Chatterjee as 'the Other', but this characterisation mainly facilitates a discussion about Hindu religion and the caste system, a widely debated issue in the 1920s and later, but not about tourism.

Modern Balinese literature, particularly that in prose form, is often preoccupied with social issues, for example caste conflict. Tourism, for which Bali is famous, started in the 1930s and began to develop rapidly in the late 1960s. Given the impact of tourism, it is perhaps natural that the social relations deriving from tourism would become a powerful source of inspiration for Balinese writers, whether writing in Indonesian or Balinese. Poetry, novels, novelettes, drama and short stories all include themes or sub-themes of tourism or social matters which are closely related to tourism's impact on society. Where Balinese and Western characters are depicted as interacting in fiction, it is usually against the background of tourism.

The main Western characters in these works are also mostly tourists and not government envoys, the crews of expeditions or traders. There are characters that are researchers (for example university students, intellectuals) but they are identified as belonging in the category of 'tourists' by the Balinese characters. This is probably related to the fact that meetings with these Westerners are depicted as taking place at art shops, places of tourist interest (such as beaches), or at public events that carry tourist appeal (such as cremations and wedding ceremonies).

Different locations, the atmosphere of the meetings, the backgrounds and the status assigned to characters also imply different motives or intentions and genres of interactions. Just as encounters through colonialism and commerce involve relations of profit, the tourist interactions of Balinese and Westerners in the works demonstrate motives ranging from general pleasure and desire for friendship and love, with specifically sexual motives being highlighted from the Balinese side. Some are stories in which the major narrative focus is Balinese-Western interaction, while others use tourism as a backdrop for other concerns, specifically issues related to cultural commercialisation and corruption.

The first short story the 1960s that focuses on Balinese-Western interaction is *My Friend Hans Schmitter* by Rasta Sindhu (*Horison* 1969, 215–7). The author, who was himself a young journalist and writer at the time, presumably met Westerners in much the way he describes one in his story. The Balinese character represented here is an unnamed first person narrator and the Western character is a German tourist, Hans Schmitter, aged 27, bearded and longhaired, a student of botany at a university in his country. Rough as he may seem in his appearance, Hans is highly educated. His travel to Bali is a part of his three-year holiday, which is more of an adventure of youth than a leisure experience. The encounter between the narrator and Hans begins at the Badung cemetery (in Denpasar), when both were watching a cremation ceremony.

The relationship between the two characters flows smoothly, starting with their mutual interest in learning each other's language: the narrator wants to learn English, while Hans wants to learn Indonesian. This same motive, so important to a Balinese who wants to enter the tourist industry or have access to the world outside Indonesia, occurs in other works. Such is the 'barter' that characterises their interaction that the relationship flourishes and Hans becomes so familiar that he is almost like a member of the narrator's family.

> I brought him home to chat. And since then Hans came frequently again to my place. Sometimes he only went home to get some Balinese candy to keep him going on his travels. Sometimes he came to my house only to borrow a shirt and trousers for the day, since on that day he wanted to do his washing at my house. Or on some days he would

> sleep over in the back room, reading. Or sometimes we'd chat on throughout the day while studying each other's language (ibid., 216).

There is nothing but the motive of language exchange to explain why the two characters should have become close so easily. Their conversations touch on all kinds of issues: Indonesia's wealth of natural beauty, its poverty, the culture of corruption, and contemporary music. For Balinese at that time Western music by people like Bob Dylan and Carlos Santana symbolised what was interesting, new and exciting about western culture. Eventually Hans blends into the Balinese scene, and the two are described as very close:

> And the atmosphere of taking snacks together was so pleasant, that none of us ever gave much thought to the fact that we were being so friendly with Hans, a Westerner from Germany, whose attendance in the midst of traditional ceremonies was something which went without notice, we had become so friendly and close (ibid.).

On leaving Hans gives the narrator his camera, and their parting is described as 'a small tragedy'.

This short story was written in 1968, a highly significant year since it is situated at the start of the rapid development of mass tourism, for which the opening of the famous Bali Beach Hotel in 1966 and the international airport Ngurah Rai in 1969 were major signposts (Picard 1990, 5). By the late 1960s Bali was receiving between 10,000 and 23,000 tourists per year from all over the world (Vickers 1989, 186; MacRae 1992, xii), while at the same time young westerners, generally known as 'hippies' in Southeast Asia, began to discover Bali. Most of them preferred to live with the Balinese and experience their culture directly, and Hans's experience, although he is not called a 'hippy', is typical of those who came to the island at this time (MacRae 1992).

Little is known about Rasta Sindhu (1943–1972) as a writer since he died at a relatively young age (of tuberculosis) and his works have not yet received a thorough study. He studied archaeology, wrote many poems and short stories, and published nationally during the 1960s, including working closely with the author Gerson Poyk. In 1969 he received a literary award from the major literary journal *Horison* for his short story *Ketika Kentongan Dipukul di Bale Banjar*, When the Wooden Drum Sounds at the Community Meeting Pavilion (*Horison*, January 1969, 27–29). This work was based around the conflicts resulting from issues of caste and the status and differences attached to each caste (Hoerip 1986, 287), but *My Friend Hans Schmitter* has nothing of the other work's preoccupation with clashes over tradition, being entirely set within the modern context of tourism.

In the late 1960s and early 1970s the most frequented tourist spots such as Sanur and Kuta, become famous for offering sexual opportunities for young Balinese men (Sabdono and Danujaya 1989, 163–7). In that period 'free sex' was associated with Westerners, and Western women in particular were regarded as much 'easier' than their Balinese counterparts. Faisal Baraas, a Muslim Balinese born in Western Bali in 1947, offers a salutary tale about the limits of such newfound sexual freedoms in his *Sanur Remains Busy*. This story was originally published in the popular magazine *Varia* (29 April 1970) and later included in the author's short stories collection *Leak*

(Baraas 1983, 47–53). Baraas is also a doctor specialising in heart disease, although in this short story he is interested in other affairs of the heart. The story describes a relationship between a Balinese man, I Wayan Sumerta, and Western woman, Joice (not Joyce). Sumerta is a university student who spends his vacation as a freelance tourist guide in order to practise his English; Joice is a university student from the United States who wants to conduct research in Bali with Karla, her lesbian lover.

The first encounter between Wayan and Joice takes place in Sanur, and in the opening scene, Wayan Sumerta demonstrates his sophistication by driving a car to the Hotel Bali Beach and is mistaken as being from the capital city, Jakarta; he thinks the hotel is *megah*, splendid. Wayan offers to assist Joice in conducting her research as a means to improve his English proficiency. She declares her wish to become a Balinese, which Wayan is pleased to hear.

The story begins with a flashback to the original meeting and romance of two years ago, anda thorough examination of that relationship; Wayan reflects on how he will meet with Joice again. As presented by Rasta Sindhu, the encounter between the Balinese and the Westerner takes place quickly and smoothly, with no cultural or linguistic hindrances. The whole relationship is very romantic, and the writer explores the details of the sexual relationship between two. Wayan sees Joice as another '...western woman who is free, open and pleasing' (ibid., 50), so he cannot understand why his sexual overtures are rejected. Once he realises that Joice has lesbian tendencies, Wayan changes the way in which he courts her. Rather than being masculine, he projects the softness of a woman. Eventually, Wayan and Joice consummate their relationship and she rejects lesbianism:

> You're not just my *guide* [English in the original], you are my doctor. You made me understand the reality of life. You have cured me of my illness. I was a victim of modern American life. Lesbianism is just like an epidemic there. Lots of women, like me, are attracted to it. And that night, you surprised me, then I realised, that this is the truth, this is what is simply natural (ibid., 51).

Wayan and Joice continue to enjoy a close sexual relationship and intend to marry. The story takes up again when they agree to meet at Sanur, the place of their first encounter. But as they go to Sanur Beach, both have serious doubts, because they have both married since their last meeting. As Wayan sits on the beach he thinks 'She'd be a complete fool to come here as we promised. Love has to give way to reality: it's too difficult for East and West to meet' (ibid., 52). Joice and her sister look down on him from the seventh floor of the Hotel Bali Beach as he goes off. 'He'd be a fool to wait for me as we agreed' thinks Joice to herself, having been told by her sister that he is dreaming down there (ibid., 52–3). Wayan Sumerta is in fact going back to his wife and child, while Joice has a husband in America.

This relationship between a Balinese and a Westerner started harmoniously, developed into a romance, but, though they were deeply in love and prepared to marry, they eventually went their separate ways. Perhaps, in spite of everything, it was just a holiday romance and their love was superficial. Here Western culture is depicted as corrupt and decadent through the naive view of lesbianism presented by the writer. The ending is in some ways cynical and the way the two women look down from their luxury hotel symbolises a vast material and cultural gap that cannot ultimately be bridged; the hope of bridging it, and Wayan's own hopes of gaining worldly prestige, come to nothing.

Romances between Balinese men and Western women are dealt with in a number of other Balinese works such as *Remembrance Monument* by Ngurah Parsua (1986, 48–56), a novelette by Gde Aryantha, *Suzan*, originally serialised in the popular women's magazine *Sarinah* in 1988. *Remembrance Monument* describes a love affair between Sudarma, a medical school drop out, and Elizabeth Yane, an Australian student holidaying in Bali. They meet at Kuta Beach, which by this time had become symbolic of up-market tourism, and the key place for contact with westerners, surpassing Sanur. Motivated by sexual desire and love, Sudarma makes the initial approach, impressed by Elizabeth's beauty.

> She's a young girl whose beauty really hits you right between the eyes. "I'm sorry, but I can speak only a little English" [English in the original] I replied awkwardly. She laughed, while I stared at her with my heart pounding. The white-skinned girl looked even more beautiful with her smile and lively laughter (ibid., 49).

If Elizabeth had not been so beautiful, Sudarma's passion and spirit might not have been so ardent. They quickly fall in love, develop a trusting and passionate relationship, and intend to marry. However, the author, Ngurah Parsua, a civil servant, seems unwilling to allow a Balinese and Westerner to unite in marriage, even though they both love one another, and Elizabeth is, conveniently, killed in a plane crash before this can happen.

In the novelette *Suzan*, a Balinese journalist, Bram, and an American detective holidaying in Bali, Suzan Hayes, form a complex relationship, but their dream to live together never materializes. Bram meets Suzan for the first time when, as a tourist, she comes to watch cremation. Bram is motivated to make an approach to Suzan both for sexual/romantic reasons and to learn English from her. Indeed one of Bram's friends sees the prime role of western women as that of an instrument for language acquisition, a view with which Bram does not necessarily agree:

> Samuh has indeed had many tourist girlfriends. He's been to Perth and Sydney. Eventually, when a white girl wanted to take him to Canada, he refused. "I don't want to be sucked in", he told me once. "I've got what's most important: facility with English."

Most of the first part of the narrative describes how Suzan's beauty radiates from her, and how much Bram is attracted to her. Suzan is attracted to Bram because he makes a suitable companion; as a detective it suits her to be with a journalist, a fellow investigator. After their initial meeting they continue their relationship in Ubud, where they finally make love:

> Her hand goes under my shirt, caressing my chest. Her fingers start to play, slowly scratching softly. "Bram", she whispers in my ear, "Are you sleeping?" I pretend to stretch for a second and then stay quiet again. Suzan kisses my neck.

At the peak of their intimacy, Bram has to accept a shocking reality – Suzan is a member of an American mafia. Her American enemy assassinates her and once again the reader witnesses the writer's apprehension in joining Balinese and Westerner characters with any permanence. They are described as having the ability to love but not to get married – either because of personal decisions or accidents.

The encounters all begin with indications that friendship is already there, and that such friendship can rapidly develop into romance. But in the end the characters have to accept the bitter reality that they have to separate. Balinese writers are simply unwilling to allow their Balinese characters to marry Western ones, and interestingly certain Western authors mirror their outlook. For example, in G. Francis's *Nyai Dasima*, Miss Dasima (1982, 225–47), marriage and a finalisation of relationships is avoided by narrative devices ranging from personal reconsideration to death. It would appear that in modern Balinese literature, a strong sense of 'East is East' and 'West is West' remains. If one compares the writings on Balinese authors with non-Balinese Indonesian ones, such as Nh Dini, Titie Said and Sunaryono Basuki, similar concerns can be detected, through the relationships between Indonesians and Westerners are commonly allowed to become more complex. Marriage does take place between their Indonesian and Westerner characters, but it does not last long, brings suffering to the Indonesian character, and ends in separation.

Encounter in Tampaksiring

The only fiction written in the Balinese language that depicts Balinese-Western interactions within a family context – and not in terms of friendship or courtship – is *Encounter in Tampaksiring* by Made Sanggra (1975, 41–63). As a whole, the nuances of the interaction in this short story are presented in a very romantic and open manner. This short story illustrates an accidental encounter between a Dutch journalist, Van Steffen, and his Balinese mother Ni Luh Kompyang, at Tampaksiring, a tourist site made popular by Indonesia's first president, Sukarno. This story begins with the visit of Queen Juliana from the Netherlands to Bali, accompanied by several journalists, including Van Steffen. His father, De Bosch had married Luh Kompyang, but they, like the characters in the other stories already mentioned, also separated; Bosch took with him his son, Van Steffen, and Luh Kompyang kept their daughter, Luh Rai, with her.

During the Dutch Queen's tour, the government's guests stayed overnight at Tampaksiring and one day the handsome Van Steffen strolls along to Luh Rai's art shop in Tampaksiring. Their encounter is romantic, without their knowing that they are in fact brother and sister, the twist to the story that is meant to surprise readers. The theme of near incest between brother and sister is a frequent one in traditional Balinese literature, featuring for example in the *Geguritan Megantaka*, which dates from the 19th century. In Sanggra's story, as in the *Megantaka*, the potential for incest is avoided, and at the end of the story the reunion of the whole family is described in a very touching fashion.

Balinese literati regard *Encounter in Tampaksiring* as one of the finest short stories in the Balinese language and won the first prize at a modern Balinese short story writing competition in 1978. The writer is a veteran of the Indonesian Revolution, who has a great interest in Balinese literature, classical and modern, and remains active in his traditional village (Suarsa 1992). In the 1990s Made Sanggra (b. 1926) was still writing for the mass media and actively participating in seminars on art and culture. In 1998, together with another Balinese writer Nyoman Manda, Sanggra received the *Rancage Award*, an honour given to writers for their literary merit or their contribution to the development of literature in their native language. Sanggra uses traditional motifs to highlight his sense of a smooth transition between older forms of Balinese literature and modern ones, and his sense of distance between Westerners and Balinese is mixed with the possibility of closing that gap.

Non-Romantic Relations

Balinese-Western interaction does not always take place in an intimate, friendly and romantic manner as depicted above. In a short story entitled *Statue* by Nyoman Manda (*Bali Post*, 7–15 November 1978, 4), a high school teacher in Gianyar, describes the tragic life of a talented sculptor called I Wayan Tamba who comes from Ubud. He hopes to earn money for his wedding by selling two statues, of Shiva and Rama, but when he takes them to an art shop they are only valued at 15,000 rupiah. Wayan Tamba is disappointed and decides that he would rather not sell them. On his way home, he takes a rest under a tree in front of the art shop to recover from his profound disappointment. As he rests a tourist greets him; they talk a while and the man, who is not mentioned by name, wants to buy Wayan's statues:

> *"Come...on"* [English in the original] the tourist offered, while indicating his room. He held the statues with great pleasure. *"Please sit down."* He felt he was being asked to sit, since the man's hand gestured towards a chair. So he sat while the foreigner laughed out loud in pleasure. The foreigner gave him 100,000 rupiah.

This sum is large enough to realise all of Wayan's dreams. The character of the tourist has two overt functions: it illustrates the possibility that Westerners are able to fully appreciate Balinese art, and fulfils the role of an instrument of fate. The sculptor's deep disappointment at his treatment in the art shop highlights his happiness in selling to the Westerner. Here the good fortune is material, illustrating an expectation, which is generally held amongst Balinese that foreigners are generally wealthy and will spontaneously bestow that wealth on selected Balinese. Seen from this perspective, Westerners they are positive instruments of fate, with their own benevolent views of Balinese people and their art.

Wayan's fate as an artist is better than that of two other artists featured in short stories with Western characters, I Wayan Rinjin in a short story by Aryantha Rinjin's *Painting* (Cork 1996, 109–20) and Pan Nukara in Made Sanggra's Balinese *Painter* (1975, 1–7). The first of these, Wayan Rinjin, a poor artist, has no other choice but to give one of his paintings to an art shop owner, I Ketut Geria, from whom he receives only a very low price. What really hurts the artist is that Anderson, an American businessman in computers, purchases his painting for US$ 750. Rinjin is there when Anderson makes the purchase and asks about the artist, yet the owner of the art shop pretends the artist is not there.

> "*It's like Walter Spies's style.* [English in original]. Who's the artist?"
> "No one knows", replied Geria in shock. He cast a sideways glance to Wayan Rinjin who was following the discussion with great seriousness.
> "He's a man from Pengosekan."
> "What's his name?"
> Geria thought for a moment. He almost didn't have the wit to lie, because he felt Rinjin's eyes boring into his back. "But I'm the proper owner of this work", he said in his heart, "I now have control of it."

The story reveals the alienation of the artist from his art, and shows the artist as an oppressed figure in the structure of the commercialised tourist industry. Pan Nukara, a painter in *Painter*, also feels the same sadness. As an artist Pan Nukara tries his best to be creative and refuses to paint commercially. A tourist guide manipulates Pan Nukara's interaction with the tourists who puts a high value on his artwork in order to get a sale, but the artist himself receives very little money from the sale.

> Oh, ...three *bendel*...at least 30,000 rupiah is what Ida Bagus just got from the westerner. Oh Lord...how much did Ida Bagus sell the painting for. I only got 4,000 rupiah, ...and I'd worked on it for seventeen days before it was yet finished. Oh, ...this is how the world is now. The clever ones who have it all worked out, the smooth tongued, are the ones who get more than those who work hard.

These stories represent critiques of Balinese culture under commercial pressure. Interestingly, it is the Balinese characters, in the form of art shop owners and guides, who are the greedy agents of capitalism, and not foreigners, and it is the Balinese artists who are oppressed by the system. The artists are only artisans, *tukang*, and are not valued for their creativity, and the Westerners involved are relatively innocent. They too are manipulated for their money, which they dispense freely, but at least they show their appreciation of these artists which the over commercialised Balinese cannot.

Disruptive Tourists

In two works by Putu Wijaya, a novel entitled *Night Falls Suddenly* (1977) and a short story *Typical* (1993), relations between Balinese and Westerners are presented in a more negative light. Balinese-Western relations are depicted as full of suspicion and conflict; romance is not possible, and any intimacy, as in the interaction between Wayan and John in *Typical*, is a brief thing in the past; in *Night Falls Suddenly*, harmony and romanticism belong on the same level as utopia.

Night Falls Suddenly illustrates the dynamics of village society in Tabanan in coming to terms with new, western, foreign, different, and critical values that are introduced by foreign tourists. One of them, David, whose origins are unclear, meets Subali, who is holding a wedding for his son. David influences Subali with new ideas to the point where he neglects such traditional village values as cooperation, *gotong-royong*. Subali is the only character in the story that can accept David in a positive manner, whereas his children and the other villagers remain suspicious of the foreigner's odd behaviour.

> Never before had any tourist visited this village, as there's nothing worth selling here. He felt a sense of foreboding. (1977, 12)

This novel is very interesting in the way that the conflict is multi-layered but focused on a villager who is so carried away by foreign influences that he ends up being expelled from his village. Because of David's influence, Subali becomes lazy about going to the temple, or taking part in village communal work. When the village community have to conduct communal work, David persuades Subali to pay a fine, *ayahan*, instead of working, and he takes Subali to the island's capital, Denpasar. This Westerner perceives the village as being poor and underdeveloped, and he sees his role as opening the villagers' mind and being a hero of progress by persuading them to abandon practices, traditions and customs that are not profitable. David wants Subali to become a reformer in his village.

Subali is frustrated because of his failure in business and has even considered transmigration, and foolishly he does just what David tells him. He often stays at home, and he never actively participates in village events. A crisis occurs when the village is carrying out a project and David takes him to Denpasar; to show off they pass by the villagers who are doing the work:

> Subali slowly approached the head of the village. He spoke so softly, but almost everybody could hear what he said. "I am really sorry for not being able to take part in the community project. I've got something to do in Denpasar, I'm ill and I have to get some medication there. Now I'd like to pay the fine." He handed over the fist full of money that David had put in his pocket. The head of the village was stunned. Everybody was stunned (1977, 81).

While in Denpasar, at David's expense, he becomes a tourist, going to Sanur and Kuta, eating at restaurants, getting drunk and chasing prostitutes. Just as Subali was becomes fully engrossed in his new life, feeling free of traditional burdens, David disappears. Left on his own, Subali reflects on what David has taught him:

> He thought it was right. He thought again about what David had said. There was a lot that he could agree with. That foreigner's words were always in his mind: he was just like his guru. For a long time he really wanted to do something in that village. Start a reform. Because it was too difficult to preserve the traditions, while daily needs were getting greater and greater. (1977, 107)

The village community rejects the reformation and due to his bad behaviour, Subali is expelled from his community membership. He is not allowed to use the village facilities: waterspout, streets and cemetery (ibid., 72). His family are also subjected to the social sanction of *sepekang*; no one may talk to them, which means the village community considers that Subali's family do not exist (ibid., 134–5); anyone found speaking to them will be fined. It ends in tragedy when Subali's wife dies and the community refuses to let the family bury her in the village cemetery.

One of the distinguishing features of Putu Wijaya's fiction is the use of shock tactics to encourage readers to re-think their attitudes toward certain accepted assumptions. Despite the fact that the author can freely criticise the village communality through David and Subali, in the end he crushes the new values that he introduces by illustrating how strongly the value of the tradition is rooted in Balinese society (Mohamad 1994, xi–xv). From this context, Putu Wijaya's *Night Falls Suddenly* may be read as a critique of progress originating in the West and which does not conform to indigenous values.

In *Typical* the source of conflict lies in the different expectations that John and Wayan have about how they should behave. Wayan and John are old friends, and when John is vacationing in Bali, Wayan accompanies him around the tourist spots. When John is leaving for his home country he invites Wayan to come to Jakarta to say good-bye to him. Wayan accepts John's invitation and he takes with him his nephew who is also his intended son in-law. John is surprised but he has no alternative but to take along the uninvited guest brought by Wayan. While in Jakarta John's intention to be with Wayan does not materialise, and because Wayan is so busy with his nephew he completely forgets to say farewell John. At the end John tries to get rid of Wayan indirectly by giving him a night bus ticket. On his way home, Wayan curses John,

> "You red devil, John. You're a typical white and you always will be. You want to be the boss and walk all over everybody and get your own way," Wayan curses.

Meanwhile, at his house, John is also complaining,

> *Oh my God*, how could anyone behave like that! He's a typical native! A typical ex-colonial subject! He's still got the mentality of a slave! (Lingard 1995, 98)

Are You Mr Wayan?

In the late 1990s, a change occurred in the representation of foreigners, which may be linked to improvements in global communications and the advent of commercial television. Wayan Suardika's story *Apakah Anda Mr Wayan?* Are You Mr Wayan? (1998) concerns the offspring of a Balinese man and a female tourist, a subject that is usually avoided. The male protagonist in the story is Wayan, a 45 year-old Balinese writer who works as cultural guide for foreign students doing fieldwork in Bali. He is depicted as a womaniser, but it is not his sexual adventures that are emphasised but the meeting between Wayan and his children resulting from such an encounter.

The story opens with the question 'Are you Mr Wayan?' put to Wayan in his house in Denpasar by a teenage American girl from Chicago, Anna Winslet. Anna introduces herself to Wayan as his daughter and tells him all about the personal relationship he had with her mother, from their first meeting in a gallery in Ubud until their one intimate encounter in his room eighteen years ago. Wayan is completely surprised by this strange girl but it is hard for him to deny her story.

> "Is it all true?" she asked, seeking confirmation of the story about me and her mother.
> "Whatever," I [Wayan] replied, still doubting.

> "Why?"
> "So many foreign women have been my friend..."
> "Not just your friend, but your lover?"
> "Yeah something like that."
> "Wow! I didn't know my father was so interesting!" she exclaimed, and then smiled, whether sincerely or in mock, or I don't know. Then she embraced me again.
> "Whoever you are, you are still my father. At least I have a father. All my life in Chicago, how I have missed you!"

Eventually, Wayan accepts Anna as his daughter and because Anna reveals her intention to become Balinese, Wayan takes her to his home village where a series of traditional Balinese ceremonies are initiated for her as if she is a newborn baby. She is quickly accepted as a member of the family, but her grandparents continue to call her a 'tourist', an identity that has been bestowed on white people since the advent of rapid tourism development in the 1970s. When they return to Denpasar, Anna invites her father to stay in a hotel at her expense but Wayan wants her to stay with him in his house, thus expressing his responsibility to his daughter. When both of them are at home, a Japanese teenager comes to see him and she asks him the same question as Anna before her: 'Are you Mr Wayan?' The story does not explore the identity of the Japanese girl nor her relationship with Wayan, but her arrival and the question she asks, strongly suggests that she too could well be Wayan's daughter from one of his other lovers.

In addition to dealing with the romantic aspect of these relationships, the story also explores issues concerning the identity of its characters, particularly the Western protagonist. Anna's identity is depicted as being rather problematic, especially as seen from the Balinese character's point of view. On the one hand, Anna is accepted as a Balinese woman; on the other, she is still considered a tourist largely because of her physical appearance, which cannot be changed. This ambivalent view is depicted in the character of the grandmother who strongly represents the conventional view of the Balinese towards Western people. As can be seen from the following interaction with her father, Anna is portrayed as strongly committed to changing her identity from American to Balinese:

> "In essence", I replied straight away, "do you want to be Balinese or American?"
> "I want to be the first one", she answered firmly.
> "That's what your grandparents want."

As a Balinese woman, she is expected to be able 'to dance and make offerings', and in fulfilling these requirements, particularly dancing, Anna is depicted in a way that allows her a promising way out. In America – as it happens – she had learned ballet and now she wishes to do 'something experimental by combining Balinese dance with ballet'. How Anna realises her intention of becoming Balinese is not explored further, but the story clearly expresses the willingness of a Westerner to become Balinese, rather than the other way around. Thus, Wayan Suardika not only shares the view of those other Balinese writers previously discussed who are unwilling to allow their Balinese and Western characters to marry, but also affirms the superiority of a Balinese identity over a Western one.

Romance and Magic

Unlike the stories discussed *above,* Putra Mada's novel *Liak Ngakak*, *The Laughing Liak* (1978), the relationship between a Balinese man and Western woman is depicted in an unusual context: namely, the tension between black and white magic. *Liak* or *léak* is a transformation performed by people who practice black magic with the *léak* taking the form of a demon, fireball, monkey or a pig. The *léak* may also appear as a bodiless head or in other terrifying forms. Stories about *léak* can be found in a large number of both modern and traditional Balinese texts, yet *Liak Ngakak* appears to be one of the most intriguing because it depicts a Westerner as its central character. Because the Western figure is depicted as seeking and practicing black magic, it reinforces an earlier image of Westerners as fire-breathing demons or white monkeys.

A belief in magic (black and white), of which sorcery, *kesaktian*, and *léak* are a part, is central to Balinese belief systems (Covarrubias 1937; Geertz 1994) and, although such beliefs date back many hundreds of years and infiltrate many aspects of Balinese life, they remain mysterious and contradictory. Ordinary people often avoid talking about *léak* because of the inherent dangers, but enthusiastically follow stories of black and white magic in the traditional performing arts.

The novel *Liak Ngakak* tells of an Australian woman, Catherine Dean, who comes to Bali to investigate *léak* or black magic after studying Indonesian for two years in Yogyakarta. Her ambition is to publish a book that will disclose all the secrets of Balinese black magic to the world at large. At a party in Denpasar, she meets Pusaka Mahendra, a Balinese man who lives in East Java and works as a sailor – as did the author. He then helps her to find a *léak* guru. Both Cathie and Pusaka are identified as adventurers, so they share common interests and personalities. From the beginning, almost nothing hinders their relationship. Cathie needs Pusaka to help her to find a *léak* guru, and in an added twist to the plot, Pusaka is drawn to Cathie's charm and beauty. Within a mere two days, under the guidance of the *léak* guru – an old woman from Sanur – and with some help from Pusaka, Cathie is already able to transform herself into a *léak* taking the shape of a pig, a bird, a monkey and a fireball. According to an acclaimed *léak* expert, Ngurah Harta, the process of learning to become a *léak* should take weeks if not months, but acknowledges that two days is still plausible depending on the guru. Here is what Cathie feels after transforming herself into a fireball, one of the most popular forms of *léak*:

> I feel that I am sitting on top of the fireball, but I don't feel any heat. Strangely, I can control the movement of the fireball as I like. I can drive it backwards, forwards, up and down, flying, gliding, diving, in short it all depends on what I want to do.

The relationship between Cathie and Pusaka develops, and during the day, they travel around Bali behaving like tourists. Pusaka and Cathie agree to avoid Sanur, Kuta, and Nusa Dua for sightseeing because hippies populate those beaches. This not only reflects the reality of the tourist world in Bali in the 1970s, the time the novel was written, but also makes a passing negative reference to the hippy phenomenon.

To help develop the cohesiveness of the plot, Pusaka and Cathie are shown on one occasion during their tour as watching a *Barong Dance*, a performance for tourists that depicts the unending war between black and white magic. During the night, while Cathie practices black magic with the guru, Pusaka sneaks a look at the *léak* lesson to assuage his curiosity, a dangerous thing to do as the *léak* guru may attack him. To protect himself, Pusaka consults his grandfather who is an expert in both black and white magic, but who has chosen only to practice the latter. After the consultation, Pusaka becomes fully convinced of the existence of *léak*, and knows more about the secrets and dangers of black and white magic. His grandfather gives him a magical dagger, *keris*, for self-protection.

The contrasting powers – black and white magic – embodied within the two characters starts to overshadow the romantic relationship between them. This unusual state of affairs forces Pusaka to ask himself whether or not he should continue loving Cathie. When his official leave finishes, Pusaka returns to Surabaya, and leaves Cathie alone in Bali. As is the case in other stories, this romantic relationship is only temporary, but there is an added frisson: Cathie dies as a *léak* when Pusaka accidentally kills her and her guru. He has actually come to save her life when both are struggling and unable to resume their human forms, because their bamboo shrines have been swept into the sea following heavy rain. These shrines are an essential element in the process of transformation from human being to *léak* and back again.

This novel was made into a film in 1981 by a Jakarta-based producer under the title *Mistik, Punahnya Rahasia Ilmu Iblis Lèak* (*Mystic, The Extinction of the Secret of Black Magic*), and was widely watched in Bali. In the novel, Pusaka kills Cathie and the guru, whereas in the film they die in a fight against a witch doctor, Pusaka's grandfather, who is supported by Pusaka and a crowd. Cathie's dream of writing a book about Balinese black magic also comes to nothing, an indication that she is merely used by the narrator to disclose certain aspects of the Balinese belief systems regarding white and black magic.

The transformation scenes are the most interesting and entertaining aspects of both the novel and the film, though the meaning of the story far transcends these mechanical details. On a surface level the novel can be read as a primer for young Balinese who do not initially believe in the existence of *léak*. Its most important point, however, is the description of *léak* and Balinese black magic through the eyes of a Western woman, which is a strategic act of representation, as it circumvents violation of the law of *ajawera*, literally the 'powerful lore that cannot be disseminated'. In practice *ajawera* people are not free to talk about certain topics such as *léak* and thus it is much more convenient for the writer to use a Westerner as the *léak*'s pupil, rather than a Balinese. In so doing the writer can easily manipulate the *ajawera* law as it applies to foreigners – should the law take effect, the Westerner will be the victim, and no Balinese will be harmed. This device is familiar within the social science literature as 'projection', whereby one's own unacceptable feelings about oneself are projected onto to other ethnicities.

The novel, however, goes much further than this since it illustrates an underlying unease about the interest in and ability of tourists to understand Balinese traditions. Cathie's death not only illustrates the dangerous consequences of practicing black magic but also reflects unwillingness on the part of Balinese writers to accept that foreigners might master one of the most specifically Balinese traditions and here lies one of the main messages of the novel. It emphasises the need for the Balinese to defend their traditions, and with them their identity, since globalization and contact with the Western world are unavoidable. The author acknowledges that it is impossible to deter external influences and in any case the Balinese need to catch up and modernise, but specific local traditions that underpin identity need to be defended at all costs, even at the expense of distraught friends, heartbroken lovers, and the occasional dead Westerner – fictional of course.

Balinese-Western Interactions

In their encounters with foreigners, the Balinese appear to be out-going, and ready to cope with change and new experiences. The friendships between Balinese people and Westerners is the vehicle of such experiences, but it holds the threat of wresting them away from their Balinenese-ness, either through becoming too sophisticated and Jakartan like the Wayan of *Sanur Remains Busy* or through a clash of values as in Putu Wijaya's work. Balinese writers never allow their Balinese characters to become assimilated by making them Westernised, either through direct adoption of values, or through marriage. With the exception of *Encounter in Tampaksiring*, all of the characters eventually separate.

Why are Balinese writers unwilling to allow Westerners a place in Balinese society? Why are they keen on killing off their Western characters when the relationships become too intimate? In order to answer these questions, we should ask why Balinese writers are interested in depicting Balinese-Western interaction in the first place.

Balinese writers from the 1960s onwards have wanted to comprehend and at the same time respond to the new phenomenon of mass tourism and its impact on their society. Mass tourism has meant that Westerners, as the first and most prominent group of tourists, have been the focus of Balinese debates about the consequences of modernisation through tourism. Picard argues that during this period an official discourse arose in which tourists were seen as a group that should be managed, beginning with anxieties about 'hippies' as a source of cultural debasement (1996, 79, 226). This official discourse may be characterised as 'us' Balinese (as well as Indonesians) versus 'them' Westerners, a re-imposition of a dualist model of identity that helped the Balinese in resolving earlier encounters.

These authors, however, also make use of this dichotomy to simultaneously criticise Balinese customs, using these Western characters as instruments to articulate views on Balinese traditions and culture. Rasta Shindu uses his character Hans Schmitter to address issues such as corruption, backwardness, and poverty. Putu Wijaya uses his character David to pit traditional values against ideas of progress, although ultimately it is the forms of 'progress' that become a problem. In these stories we move between Indonesian, English and Balinese, with even some of the stories in Indonesian making use of Balinese language at strategic points. Rasta Sindhu's story, for example shows how the meeting of languages can create its own harmony which is quite outside any official views of 'hippies' or Western culture, while Sanggra's stories also display great depth and humanity, a consequence of the Balinese having had long-term contact with Westerners.

Figure 6.1 Satellite dish behind a traditional village courtyard

Within Balinese culture, criticism, or even compliments from other Balinese are not considered polite or a normal part of discourse. People are taught that one should not comment on oneself, as can be seen from a line in the well-known traditional poem *Geguritan Basur*: 'Do not judge yourself, let others do it for you', '*eda ngaden awak bisa, depang anake ngadanin*'. Using 'the other' to present such views is much more meaningful and appropriate, so foreigners – who are considered outspoken, if not gauche – are a vehicle for saying what Balinese characters cannot say. It is important then that the English-speaking Westerners are used in these works because they are at a greater cultural distance than members of other Asian cultures. Thus despite the recent high number of Japanese tourists who have come to Bali – including those who have married Balinese – there are as yet no works of fiction depicting Japanese characters. Likewise marriages between Westerners and Balinese that have endured are outside the frame of reference of these stories; these depictions of 'reality' do not necessarily accord with the 'realities' of Balinese life.

Balinese-Western interactions in fiction are not about acculturation or even marriage, but are part of a process by which Balinese intellectuals try to comprehend what is good about Balinese culture and what is good for the Balinese. The objective is not to understand, let alone to promote Western culture, because their authority to depict Western values through their characters does not necessarily bear close scrutiny, such as Baraas's simplistic view of lesbianism. In these works of fiction the characters are very artificial, and they serve as vehicles for the writers to express their personal and social ideas and experiences overtly.

Balinese authors use Balinese-Western interactions in fiction as a device to help their readers comprehend and interpret their own culture, as opposed to an in depth analysis of what it means to be Western. These approaches would appear to accord with Miguel Covarrubias's observation that the Balinese are adept at adapting foreign ideas to their own culture and that they have shown an unusual logic and intelligent power of assimilation (1956, 403). These Balinese writers from the 1960s to the 1990s seem to be aware that 'assimilation' is probably too simplistic a way of seeing the process. The need for change that comes from the impact of the tourism and the onset of globalization is shown in these works of fiction as being evaluated in various ways, using various mechanisms to emphasise various features of Balinese culture.

Balinese-Western interactions may utilise realties resulting from the development of tourism in Bali, but in works of fiction these are handled in specific ways that illustrate Bali's cultural dynamism. The works discussed above all include some form of distancing: either the need for separation of characters who may otherwise marry, or the portrayal of Westerners as remote and even negative. This depiction involves a certain amount of stereotyping of Western culture, what we might call 'Occidentalism', in accordance with the ideal that the Balinese are open in their socialisation with Westerners, but are not easily westernised. These stories of global-local encounters simultaneously alarm the reader with the potential threats posed by foreigners in one's midst and reassures them though the resilience of Balinese culture, and reflect a more widespread mood about how the islanders are to cope with this latest wave of interesting, but potentially dangerous outsiders.

Chapter 7

World Heritage as Globalization

The acquisition of UNESCO's coveted World Heritage label is usually assumed to be an honour for the people in the vicinity of the site designated as such, and only very rarely do complications arise in which local stakeholders become opponents and the nomination process is brought to a halt. Given that UNESCO defines 'Cultural Heritage' as comprising a monument, group of buildings or site of historical, aesthetic, archaeological, scientific, ethnological or anthropological value, then one would assume that an island as renowned in Bali would have sites suitable for inscription on to the World Heritage list, but when the Indonesian government proposed to nominate Bali's most distinguished temple complex, *Pura Besakih*, for this accolade there was stiff local resistance. Contrary to the widespread expectation that local people would be delighted with a World Heritage nomination, key stakeholders in the island opposed the process and forced the suspension of the nomination.

Could it have been that after more than a quarter of a century that the convention that created the World Heritage award had lost some of its uniqueness or were there other interests at stake that undermined the perceived benefits and costs of this designation? As this chapter will argue, it was not dissatisfaction with UNESCO's label per se that provoked such resistance, but the manner in which the nomination was handled by the Indonesian authorities. There were widespread fears that the nomination would compromise the religious standing of the temple and that the islanders would lose control over key aspects of its management, a concern compounded at the time by worries about tourism's impact on Balinese culture. The situation was further complicated because as the mother temple in Bali Besakih was also the site of a great deal of local rivalry. Bali's clans have rights and obligations relating to the site and there are competing claims of precedence among the most ritually revered tier of Balinese society, the so-called priestly caste who bear the title Ida Bagus.

A further complicating factor was the ongoing reform of Indonesia's political system, which also had an impact on the dispute over Besakih's World Heritage nomination. Since the fall of Suharto in 1998, successive Indonesian premiers have tried to reverse the centralist tendencies of the New Order era and to devolve power to the regions. A Law on Regional Autonomy (UU Otonomi Daerah), for example, was promulgated in May 1999, but only implemented on January 2001, but it was not the provinces to which power was devolved but to the districts (*kabupaten*) and municipal authorities (*kota*) (Picard 2003, 115). In Bali the boundaries of the today's districts were established during colonial times and are based on the island's old kingdoms, and thus the term regencies is also used to refer to districts. After independence the Dutch East Indies became Indonesia and the law setting up Bali as a province was enacted in 1958 (Stuart-Fox 2002, 316). At the time Bali was unique in that the provincial boundaries were equated with the island's boundaries, a distinction not granted to any other island in Indonesia. Reform, however, has created a new set of worries in that devolution to the districts will dissolve the island's unity, re-awaking fears that the enmities that divided the small kingdoms of pre-colonial Bali will return. If the island is not united, then how should one view any attempt to restart the World Heritage debate since the honour of inscription would not apply to the whole island but only to a portion of it? These are especially pressing concerns in the light of the decision to press ahead with nominations for three other sites in Bali: Taman Ayun Temple, the Jatiluwih rice terrace and the Petanu valley archaeological remains.

Despite their worldwide significance very little research has been conducted on the processes by which World Heritage Sites are nominated and accepted. This is not to say that research on World Heritage Sites has never been undertaken because there are a number works (for example Shackley 1998) that have set out to do this, but only a small number of academic papers (for example Hitchcock 2002; Smith 2002) have analysed the manner in which World Heritage Sites are inscribed. Accounts of failed nominations are even harder to find, but the importance of understanding how this may happen is underscored in a paper by van der Aa, Groote and Huigen (2005). These authors documented a case in the Wadden Sea where inscription costs were seen as outweighing the benefits for those who lived and worked in cultural landscapes being nominated. The paper points out that local opposition cannot be ignored, especially when the perception is that certain stakeholders have more to lose than to gain through being on the World Heritage List.

The Temple of Besakih

Situated on the east of the island at 1000m above sea level Pura Besakih, the largest and most important temple complex in Bali, comprises no less than eighty-six structures, most of which are gathered around Penataran Agung, the most significant of them all. The term pura is often translated as 'temple', but it is not a temple in the usual sense of the word and is instead an area of sacred ground, where the gods are venerated. The term 'temple' also does not fully express the complexity of Besakih's religious buildings since there are three different kinds of structures: twenty two state shrines, fifty three descent group shrines, eleven locality shrines (Stuart-Fox 2002; Geertz 2004, 286). Gunung Agung (3,142m), Bali's highest volcano, provides a backdrop for the temple complex and, in accordance with Balinese beliefs, is associated with kaja, the abode of gods and deified ancestors personifying order. Kaja is the upper end of a continuum, the opposite of which goes downwards towards the sea, kelod, in the direction of demons who signify disorder (Hobart, Ramseyer and Leeman 1996). There are many thousands of pura on Bali and most villages and urban wards have at least three, and often more, each one serving a specific purpose.

The foundation of Besakih in the 9th century AD is associated with Rsi Markandeya, one of the great Hindu Rsi, and in Balinese tradition he is said to have originated in India (Stuart-Fox 2002, 261), though it is possible that he was Javanese. According to the legend, he buried a pot filled with water in the ground containing *Pancadathu*, the five elements of gold, silver, copper, iron, bronze, together with a precious jewel. The site was given the name Basuki, which is said to mean 'prosperity', because the ritual was conducted to ensure the success of the venture (ibid.). In 11th century, another Javanese proselytiser, Mpu Kuturan, arranged the space, order and architectural style of the temple (Dinas Kebudayaan Bali 1999, I-3). Following incorporation into Majapahit's sphere of influence, Besakih underwent some further changes and this continued after the fall of Majapahit in Java under the Kingdom of Klungkung. It was during the latter period that a royal decree, Raja Purana, was issued saying that it was the king who held the highest authority and thus Besakih was under his jurisdiction. It was at this time that Besakih became a state temple, but it remains unclear whether the king established it as such or simply appropriated what already existed (Reuter 2002, 377 fn 10).

After independence in 1950, overall authority devolved to the republican government and its local representative, the Governor of Bali. In theory, the Governor was responsible for the maintenance of the site, but in practice it was the clans who exercised control over their respective temples, while spiritual authority remained in the hands of the priests. The Governor soon divested himself of this responsibility when he dissolved the Panitia Pembinaan Besakih, the governing body that came under the provincial government, in accordance with the view that matters concerned with religion should be organised by each religion's supreme council rather than by government (Stuart-Fox 2002, 316–7). He handed over responsibility for the temple to the Parisada Hindu Dharma and demanded more intensive management of temple-owned lands, *laba pura* (ibid.). The temple was not, however, disentangled from government and a rather more complex arrangement emerged in 1975 when a Masterplan was drawn up for Besakih and its immediate environment. The Directorate of Culture appears to have been behind this initiative and the Director-General at the time was Ida Bagus Mantra, a Balinese (ibid., 318), who served as Governor between 1978–1988.

Lack of clarity with regard to the functioning of the temple has been an ongoing problem in post-independence Indonesia. A further complicating factor is that Besakih has long been one of the island's tourist icons with visits by international tourists dating back to the 1930s. As domestic tourism grew in the late Suharto period (mid 1980s and 1990s), Pura Besakih began to attract tourists from all over Indonesia. Many of these external visitors were Hindus from other islands, who might more accurately be described as pilgrims, though no hard and fast line may be drawn between them and tourists. To impress upon tourists the significance of the site, visitors are subjected to more regulations and restrictions than almost anywhere else in Bali. There are guidelines on what to wear, as well as a signing in book for independent travellers, and many parts of the complex are closed to non-Balinese. The combination of visitation rules and rising entrance fees for tourists, as well as disputes with tour operators and local guides, keen to defend their territorial rights, has led to a gradual decline in Besakih as a tourist destination.

Another factor to take into account has been the growing role of the cluster of villages that are located near Besakih and are known as Desa Adat Besakih. Until recently their management role within the temple was unclear and indeed it would have been highly unusual for such a small institution to take responsibility for a complex as prominent as Besakih. Nonetheless, as this chapter will demonstrate the collection of villages has emerged as an increasingly important stakeholder, but they are unlikely to be given overall strategic authority for the site.

Figure 7.1 Worshippers at *Pura Besakih*, Bali's largest temple complex

World Heritage in Indonesia

The Convention Concerning the Protection of the World Cultural and Natural Heritage has been signed by more than 150 states, known as the 'states parties', and Indonesia is one of the signatories. UNESCO's remit is to define and conserve the world's heritage by listing sites whose outstanding values should be safeguarded for all humanity and to ensure their preservation through closer co-operation among the states parties to the Convention. By signing the Convention, each nation promises to conserve all sites located within its borders and not just those designated as World Heritage Sites. In theory the international community as a whole is responsible for the preservation of these sites for future generations. UNESCO's aim is to encourage states to sign the Convention to help conserve their own natural and cultural heritage. Adopted by the General Conference of UNESCO in 1972, the Convention regards 'Cultural Heritage' as comprising a monument, group of buildings or site of historical, aesthetic, archaeological, scientific, ethnological or anthropological value. Since Besakih is considered to be a cultural heritage site, the natural heritage aspects of the convention are not discussed here.

One of the main tasks of UNESCO's World Heritage Centre is to invite signatories to the Convention, the states parties, to nominate sites within their national territory for inclusion on the World Heritage List. It is therefore the national governments of each participating nation that have responsibility for putting forward sites for the honour of being World Heritage Sites and as this chapter documents Indonesia has been active in this respect. It is difficult to be precise about how these invitations are taken up since they are often preceded by dialogue between the nominating countries and UNESCO officials, with the latter being well informed and sometimes having a professional interest in the sites being discussed. In the case of Besakih two senior UNESCO officials had visited the site by 2000, and were actively encouraging the nomination process.

It is partly because of this that there is a widespread perception in intellectual circles in Indonesia that UNESCO invites governments to submit specific World Heritage applications. Formally, however, the invitation from UNESCO is merely an open invitation to submit bids and it is not UNESCO that selects the sites that will be submitted, but the national governments themselves. The difference between an open invitation and a specific one may appear to be splitting hairs, but in the Indonesian context this is really quite significant. Newspaper and magazine articles on the subject often present UNESCO's position as actively seeking out sites to include. For example, if one reads *Tempo*'s coverage of the Besakih story in September 1990 the following sentence appears at the outset:

> Mula-mula ada permintaan dari UNESCO kepada pemerintah Indonesia agar mengusulkan warisan budaya apa saja yang perlu ditingkatkan menjadi "warisan budaya dunia".
>
> Initially there was a request from UNESCO to the Indonesian government to propose whatever cultural heritage that needed to be elevated to become "World Cultural Heritage" (*Tempo*, 29 September 1990).

However, despite being a willing participant in UNESCO's programme, it has taken time for Indonesia's nominations to bear fruit and thus the earliest sites only date from the 1990s, the first being Borobudur (1991), Prambanan (1991), Komodo (1991) and Lorenta Papua (1991). Being convention signatories, Indonesia has been actively nominating its cultural and natural heritage to be adopted as World Heritage. The subsequent nominations were widely welcomed, and thus it came as a surprise to the Indonesian government that when it was proposed that Besakih should be included there was widespread resistance. Nominating Besakih was moreover in line with Indonesia's state code of Pancasila, in which gives equal weight to all the country's recognised religions, an important consideration given the earlier nomination of the Mesjid Demak in Central Java, associated with the country's Muslim population. The Minister of Culture and Tourism at the time, I Gde Ardika, also wanted to convey to the world Indonesia's commitment to equal status with the majority Muslim population of Bali's Hindu minority, especially with regard to heritage conservation. In addition, there were concerns about provincial equity, especially with regard to provinces outside Java. It can also be said that Indonesia's national priorities were influenced by an awareness of moves by other Asian countries to safeguard their cultural diversity. Underpinning this were the economic imperatives of maintaining and developing Indonesia's tourism industry.

Besakih's Nomination and Rejection

The nomination of Besakih has always been of intense public interest, not least because of the existence of comparatively well-educated stakeholders with an interest in their constitutional religious position. The debate, however, was often tangled with overlapping spheres of interest, but for the sake of clarification there were four main stakeholders – the national government, the provincial government, the religious bodies and the intelligentsia – with the same personnel often being simultaneously members of more than one of these.

Proposals to nominate Besakih as a World Heritage Site have emerged three times in one decade, but have always ended up being rejected by Balinese of various persuasions. On the first two occasions the proposals were rejected without any significant voice in favour, but the last attempt provoked a heated debate on the pros and cons of nomination among the Balinese intelligentsia, though those who opposed the idea received the lion's share of media coverage partly as a result of being in opposition to the government backed initiative.

The first step in the nomination process occurred when the representative of the national government, the Coordinating Ministry of Public Welfare convened a Working Group on cultural and natural heritage in 1990. But, the Hindu Council Parisadha, representing Bali's religious interests, rejected the proposal immediately through its spokesperson, I Ketut Wiana. He objected in particular to the term *warisan* (heritage) because it appeared as if the people had abandoned Besakih. The Council and the intelligentsia opposed to the nomination assumed that sites such as Borobudur, with which they were familiar, were representative of World

Heritage Sites in general; they did not want Besakih to be treated like Borobudur where ritual activities had been banned (*Tempo*, 29 September 1990). They were, however, unaware of examples of what are sometimes called 'living' World Heritage Sites, which are actively lived in and used, often for ritual purposes, since there were no examples close to hand in Indonesia. The rejection by the Hindu Council is understandable, but represents a misreading of UNESCO's mission. The debate also took place at a time when information flowed less freely, not only because of the authoritarian government of the time, but also access to the media was not as widespread as in the current century. In the final outcome Besakih was removed from the Working Group's proposal and the Governor of Bali, Ida Bagus Oka, siding with Hindu interests and the intelligentsia, announced the rejection in the regional assembly in Bali (*Bali Post*, 4 October 1990).

Controversy resurged two years later when the national government issued law number 5/1992 on Cagar Budaya (heritage conservation) which would make it possible for Besakih and other temples to be listed as national heritage. Temples nominated in this way would thus join 143 other sites and temples that had already been declared as national heritage (*Kompas*, 12 January 1993). A deputation representing the Indonesian Hindu Intellectual Forum (FCHI or Forum Cendekiawan Hindu Indonesia) went to see Minister of Education and Culture, Fuad Hassan, to persuade the government not to include Besakih as either national heritage or World Heritage. For the second time national government had to back down and the Minister of Education and Culture announced on central television the cancellation of Besakih's nomination.

Despite these two rejections, enthusiasm for nominating Besakih did not wane in national government circles and a third proposal appeared in 2001, though this time it was a Balinese Minister supported by provincial government and some intellectuals who took a leading role. The minister in question was I Gde Ardika and his argument was that Indonesia wanted to show to the world how well it cared for its heritage (*Bali Post*, 5 October 2001). Part of the stimulus for this new initiative seems to have been an international conference on Cultural Heritage Conservation at the Grand Bali Beach Hotel in July 2000, which had reignited interest in the need to manage Besakih. Prior to the conference, a feasibility study for a master plan of Besakih carried out by Bali Cultural Affairs (*Dinas Kebudayaan*) under the support of a private international consultant. To follow it up, a socialising or embedding team for World Heritage was formed and headed by Prof Dr Gde Parimartha who was the head of Cultural and Tourism Research at Udayana University. The team comprised researchers from the university as well as other members of the intelligentsia and was also responsible for drafting the proposals to nominate other sites, namely Taman Ayun and Jatiluwih rice terrace.

While the team engaged in 'socialising' the proposal, the national government announced a new plan on 1 October 2001 at the national assembly in Jakarta and an opposition materialised instantly with memories of what had happened before. As the debate heated up, the opposition tried to persuade the minister, I Gde Ardika, to resign and accused him of lacking sensitivity, as a Balinese, to Balinese feelings.

> Lebih baik Pak Ardika mundur saja. Kebijakannya mengusulkan Pura Besakih sebagai WBD (Warisan Budaya Dunia) menandakan beliau tak memahami karakteristik orang Bali, dan pariwisata Bali itu sendiri.
>
> It is better that Mr Ardika steps back. His policy to nominate Besakih Temple as a World Heritage Site indicates that he does not understand Balinese character, and Bali's tourism as well (*Bali Post*, 4 October 2001).

The minister responded by reassuring his critics that even if Besakih were to become a World Heritage Site then there would be no attendant loss of local control. He further argued that those who opposed him misunderstood his intentions or lacked sufficient information on the nomination process.

The debate became so acrimonious that Ardika announced that the nomination would be halted, but by December he had recovered his nerve sufficiently to reaffirm his commitment to nominating Besakih (*Bali Post*, 15 December 2001). There was a renewed round of protests and the nomination was postponed, with many of the opposition, including the head of the intelligentsia's rescue team for Besakih, Made Kembar Kerepun, remaining sceptical about what postponement actually meant (*Bali Post*, 6 October 2001; *Sarad*, 22 January 2002).

While all this was going on, the nomination of temple of Taman Ayun was proceeding smoothly under the leadership of the senior custodian of Mengwi Palace. In this case, clear lines of ownership, in contrast to Besakih's tangled pattern of control, freed up the decision making process (*Denpost*, 10 August 2001). But, despite a comparatively straightforward process of nomination, Taman Ayun, which along with many other temples in Bali had gained the status of national heritage (*cagar budaya*), was still waiting for agreement by UNESCO in 2004.

A complicating factor was that the Minister of Culture and Tourism's position was not completely isolated and there was some support from Balinese intellectuals. Prof Putra Agung, for example, was quoted as saying that the rejection of World Heritage Site nomination should not be an emotional issue and for him all that was needed was to think critically. As long as the holiness of the temple was guaranteed and the ownership remained in the hands of Hindus, World Heritage status, according to these intellectuals, would not be problematic (*Bali Post*, 6 October 2001).

Eventually, however, the provincial government put its weight behind the proposal and, according to Ida Bagus Pangdjaya, head of Bali Cultural Affairs at that time, the idea to nominate Besakih came from three leading Hindu intellectuals: Ketut Wiana (despite his earlier opposition), the late Prof Dr I Gusti Ngurah Bagus, and Ir I Nyoman Gelebet (*Sarad*, 22 January 2002, 29). Despite fears that it was central government that was orchestrating the nomination, it would appear that the initial step came from Bali and not Jakarta. Advancing his arguments in the same issue of Sarad as Pangdjaya, Nyoman Gelebet also lent support to the development of Besakih, which had become 'chaotic and wild' (ibid.). Gelebet moreover criticises those who rejected the proposal as being morally hypocritical, as they failed to encourage the investigation of temple assets, notably land, though not specifically those of Besakih.

Broadly speaking, those in favour of nominating Besakih as a World Heritage Site tended to advance the following arguments. First, they maintained that Besakih was a large and chaotic site and that expert help was needed to develop it and manage it efficiently. Second, they saw UNESCO as the arbiter of the best archaeological and conservation procedures and wanted to ensure the temple was handled professionally. Third, they saw World Heritage status as an international hallmark of quality that would raise the temple's profile and thereby attract more visitors. Finally, they were persuaded that not only was UNESCO a source of finance, but it was also a route to other sources of international support.

In order to alleviate local fears it was suggested that in the event of a successful nomination Besakih would remain under local control for care and maintenance (*Bali Post*, 5 October 2001). This was not just a hollow political gesture since there was a local precedent for involving both the local community and the outside world in Besakih's conservation. For example, when an earthquake damaged the temple on 21 January 1917, funds for reconstruction were supplied not only by the Balinese, but also by the colonial government. Ordinary people were asked to make a donation between 100–500 kepeng, while the Dutch provided 25.000 guilders, of which 1.000 guilders alone came from Princess Wilhelmina (Stuart-Fox 2002, 302). The central government was keen to develop Besakih as a World Heritage Site in order to advertise Indonesia's commitment to heritage conservation, whereas local government and certain Balinese intellectuals were more concerned with raising money for restoration and in attracting support from conservation specialists. Lofty ideals can be detected in the government's position, which contrasts markedly with local concerns about ruined buildings, chaotic planning and efficient waste disposal.

Another argument that was raised in favour of nominating Besakih was the potential revenue to be derived from cultural tourism. Pangdjaya, representing provincial government, argued that Bali would receive more 'cultural tourists' if Besakih was listed as a World Heritage Site (*Bali Post*, 18 December 2001). This was a somewhat optimistic perspective since Besakih was already experiencing some strains in its relationship with tourism. There were, for example, almost 300 local guides in Besakih, of whom 150 were thought to lack the appropriate guiding certificates (*Bali Post*, 11 May 2002). A further complication was the additional levies placed on tourists and operators, who were charged before they entered the temple, which in the case of buses amounted to Rp 30,000 on top of the parking fee. This unsatisfactory situation had still not been remedied by mid 2006 when the Head of the Guides Association called on tour companies and the public to report the charging of illegal fees. As reported in the newspaper, *Bisnis Bali*, Sukadana, the head of the official guides argued that the coercion of visitors by the euphemistically called 'special guides' should be stopped by educating the perpetrators, issuing of formal reprimands, and by threatening to revoke their licenses (*Bali Update*, 21 May 2006). A combination of these factors has led to the waning of Besakih's popularity as a tourist destination and most operators have dropped it from their itineraries.

Members of the intelligentsia opposed to the nomination gathered around the outspoken Kembar Kerepun who rapidly emerged as the leader of the Besakih Temple Rescue Team (Tim Penyelamat Pura Besakih). In his article in *Sarad*, Kerepun pointed out that the proposal to nominate Besakih would anger many parties and he gave four reasons for rejecting it. First, hardly had the debate calmed down following the demise of the first proposal to nominate Besakih, when a new proposal was announced publicly. Second, he argued that the timing did not take account of the sensitivities of the international climate, notably the negative sentiment associated with the attacks on the World Trade Centre, which Bali was helping to resolve. Third, the proposal had been put forward by a Minister of State, I Gde Ardika, who was of Balinese ethnic origin and who was charged with creating a stable climate for Bali's tourism industry. Finally, as a minister he should have elaborated the pros and cons of the argument and not simply come out in favour of those who were behind the nomination.

To support his calls for rejection, Kembar Kerepun quoted two articles from UNESCO and homed in on what he perceived as ambiguities in the Convention. The first one is on 'defining our heritage' and it states that 'by signing the Convention, each country pledges to conserve the sites situated on its territory, some of which may be recognised as World Heritage'. The second is article 4, Concerning the Protection of the World Cultural and Natural Heritage, which maintains that "Each state party to this convention recognises that the duty of ensuring the identification, protection, conservation and transmission to future generations of the cultural and natural heritage referred to in articles 1 and 2 and situated on its territory, belongs primarily to that State". Kerepun argued that Balinese Hindus would have difficulty in accepting these statements because they would have to hand over the protection and conservation of their temples to institutions that were non-Hindu (i.e. the institutions of the Indonesian state). In a rhetorical flourish he asked whether or not Balinese Hindus had made promises to themselves or to God to look after Besakih temple.

As the debate intensified the Governor of Bali, Dewa Beratha, met face to face with the Besakih Rescue Team and made three points. First, he said that he personally and as Governor of Bali never proposed Besakih to be a World Heritage Site, thereby distancing his stance from that of the national government. Second, he announced his decision to disband the socializing team working on Besakih's nomination. Third, he also let it be known that he would soon issue a decree to protect temples and other holy areas in Bali (*Nusa*, 14 August 2001). The Governor's position clearly demonstrated that the Besakih Rescue Team had prevailed, but after making several blunders in the media with his stop-go proposals to nominate Besakih, Minister Ardika finally announced its postponement (*Bali Post*, 23 December 2001).

This debate had a direct impact of the villages close to Besakih, but they hardly participated in the discussions. Despite their low profile, however, the villagers gradually acquired some better-defined responsibilities such as the power to grant permission to use public facilities for rituals and other related observances. They did not, however, always utilise these new powers effectively within the context of tourism and inadvertently contributed to the problems. In addition, some of the restrictions imposed on the temple impinged on the effectiveness of the villagers in working with tourists, an especially important consideration given their growing dependence on the industry. As the number of tourists dwindled, the proportion of traders to visitors increased and their business dealings became more desperate but this was often misunderstood and sometimes resented by foreigners.

The discussions continued, however, without much input from the villagers and in addition to fears that the central government would take control of key aspects of the temple complex should it be made into either a national monument or World Heritage Site, there were also concerns about the role played by Jakarta in the island's tourism industry. While the Balinese have retained a substantial stake in their tourism industry, particularly with regard to small-scale hotels, shops and restaurants, many of the larger hotels and resort-style developments, however, have been developed with funding from Jakarta and abroad. This inward investment was also encouraged by a former governor of Bali, Ida Bagus Oka (1988 to 1998) who was widely accused, though some think this is unfair, of selling out to foreign interests. There have also been persistent rumours that Jakarta based conglomerates associated with the family of former president Suharto have huge investments in upmarket tourism in Bali (Aditjondro 1995). These stories appeared to be vindicated by the publication of the results of an investigation into Suharto's assets by the American magazine *Time* (Colmey and Liebhold 1999). The publication alleges that over thirty years the Suharto family accumulated $2.2 billion in hotel and tourism assets, though their current holdings are much diminished.

The Balinese have long maintained an ambivalent view on temples as tourist attractions (Rata 1996, p. 359). Since the beginning of mass tourism development back in the 1970s, temples became one of the main draws in tourism in Bali. But, as tourism numbers rose sharply in the mid 1990s, the issue of temple visitation became more vexed. Temples became central to the discourse on tourism and concerns about the desecration of religious sites start to feature prominently in the local media and other public forums. As the mother temple, Besakih naturally became the focus of these kinds of debates and one of the solutions was to increase the numbers of areas, such as the inner sanctums of temples, which were off limits to tourists. As a result of these concerns, Besakih became an especially sensitive issue and this also had an impact on reactions to the proposed World Heritage nomination.

World Heritage as Globalization

Bali's failed World Heritage nomination is not unique in the history of UNESCO's Convention, but it is certainly quite rare. Over thirty years have elapsed since the adoption of the World Heritage Convention and, as this case study demonstrates, it can longer be assumed that inscription is an automatic honour for local populations and is a useful instrument for organisations concerned with tourism and conservation. This is not to say that the Balinese did not value their site because in fact they valued it so much that they were unwilling to relinquish aspects of its management to the national government. Lack of clarity with regard to the functioning of the temple after it had become a World Heritage Site lay at the heart of the dispute. In particular, there were also fears that the religious significance of the site would be compromised as responsibility was devolved to non-Hindu institutions, an especially sensitive issue in a country where Hindus comprise a small minority. To those opposed to the nomination, the use of the term 'heritage' implied something that was no longer used and indicates that more needs to be done to make UNESCO's message more understandable. The debate also touched on the sensitive status of Hinduism in Indonesia as a religion with fears that the nomination would turn the temple into a mere cultural or national as opposed to religious edifice. The fact that the Minister of Culture and Tourism was a minister of state and of a lower status than the Minister of Religion, who is a full minister, was doubtless a consideration.

The debates about Besakih also occurred against a backdrop of major structural change in Indonesia, which conceivably contributed to the tensions surrounding the World Heritage nomination. The devolution of powers from the centre to the periphery, which is widely hailed as a positive move, raised fears on the island about its cultural integrity and the possibility of a return to the rivalry that existed in pre-colonial times. An attendant feature of decentralization has been the raising of ethnic consciousness as a way of mobilising local people to express their concerns to the national government. Temples in particular are caught up in Balinese conceptions of identity and as the mother temple Besakih has come to embody and symbolise what it means to be Balinese and Hindu in Indonesia.

A complicating factor is that Besakih has long been and remains an important focal point not only for Indonesian Hindus in general and Balinese Hindus in particular, but also for the clan groupings that make up the ethnic Balinese. These clan groups continue to exercise certain rights over Besakih and this diffuses ownership in Besakih, making a nomination for World Heritage status difficult, if not impossible. To complicate matters further, Besakih has become something of a political vehicle, whether the issues are about ritual, development programmes or the devolution of government, for expressing power struggles with the centre. The rejection of Besakih's nominations as World Heritage Sites by two governors was not undertaken simply to protect Besakih but to win support from the public. Yet despite assurances that the appropriate regulations would be imposed to protect Besakih, the temples and complex buildings remain in poor shape and the site continues to be managed erratically.

After three failed attempts to establish Besakih as a World Heritage Site it would appear unlikely that there will be any further nominations. But a future nomination cannot be ruled out entirely since issues of cultural precedence might have a bearing on the matter. If the temple of Taman Ayun is successfully nominated and the owner is seen to remain in full control, then pressure to nominate the most important temple of all in Bali may resurface again.

Chapter 8

Bali During the Asian Crisis

The onset of the Asian Crisis that followed the flotation of the Thai Baht on 2 July 1997, signalled the end of an interim period in Indonesia, between the classic New Order era and the ending of the Cold War (*c*. 1989) and the period of reform. Known as the 'opening', *Keterbukaan*, it became more generally known in Bali as the Era of Globalization since the idea of globalization was seen as a new and more complex perspective on development in Indonesia, which was no longer the sole property of the state (Vickers 2003, 24). It was also in this period that the Balinese intelligentsia's earlier concerns about tourism's impact on the Balinese way of life had begun to be supplanted by an equally compelling worry about the effects of globalization, of which tourism was but a part, albeit an important one.

During the Crisis the island was not as peaceful as it appeared to be on the surface, but nonetheless managed to retain a relatively stable tourism industry. The island's comparative peacefulness is attributable to various intersecting interlocutors: Bali's well established international identity, the all-important tourism, the vested interests of the Chinese and other powerful Indonesian investors. These outside interests are also closely interwoven into the island's economy via well-connected islanders. These interest groups were also able to persuade the media that Bali remained safe, despite the international attention focussed on Indonesia's troubles between 1997 and 1999. The prevailing view that security is a prerequisite for the maintenance of a successful tourism industry needs to be qualified, but it remains unclear whether or not any general conclusions can be drawn since Bali remains an island with many special attributes.

Tourism and Crises

A common theme within the literature of tourism is that political stability is a precondition for the prosperity of tourism. There is a widespread view among tourism analysts that international visitors are very concerned about their personal safety (Edgel 1990, 119) and that '...tourism can only thrive under peaceful conditions' (Pizam and Mansfield 1996, 2). Tourism is perceived as being particularly vulnerable to international threats such as terrorism (Richter and Waugh 1986, 238), though analysts recognise that it may be impossible to completely isolate tourists from the effects of international turbulence (Hall and O'Sullivan 1996, 120). One of the most widely cited cases of the effect of international strife on leisure travel is that of the Gulf War in 1991. The downturn that accompanied the outbreak of hostilities had an impact not only on the area immediately surrounding the strife, but on international tourism generally. Indonesia, for example, was among those effected by the war, though it was located a great distance from the scene of conflict. Tourist arrivals in Indonesia tumbled in the first half of 1991 despite its designation as 'Visit Indonesia Year', part of an ASEAN-wide tourism promotion strategy (Hitchcock, King and Parnwell 1993, 4).

In view of tourism's sensitivity, it is also widely held, particularly by tourism promotion boards, that the press has a particular role to play in helping alleviate the fears of travellers. In this respect the media is seen as being a major force in the creation of images of safety and political stability in destination regions (Hall and O'Sullivan 1996, 107). Not only are obvious threats to tourism such as the press coverage of terrorism seen as a cause for alarm, but so is negative reporting in general. Following the onset of the Asian monetary crisis in 1997, for example, Thailand became increasingly alarmed about the future of its tourism industry in the wake of the poor publicity and sought to counter the flood of bad news by the positive promotion of the country as a cost-effective destination (Higham 2000, 133). Thailand's use of tourism to simultaneously boost its image and offset its budgetary deficit at a time of crisis has been covered in the professional literature, though it certainly merits further investigation.

Political instability combined with the spread of the financial crisis to Korea and Japan, important sources on inbound tourists for Southeast Asia, brought about a sharp decline in visitor arrivals. So severe was the recession that Malaysian Airlines, along with Philippine Airways and Thai Airways, deferred deliveries on aircraft, mostly the long haul 747–400s and 777s that provide a lifeline for the region's tourism industry. The turmoil was particularly marked in the region's tourism sector because of specific local factors linked to hotel development. Strong investment in real estate had been a feature of the rapid growth that preceded the crisis, and hotels in particular were rated as among the most prestigious properties to own. Generous supplies of capital fuelled a hotel building boom, and, though the effects were most visible in Jakarta, Kuala Lumpur and Surabaya, there was also an upsurge in resort development. The downturn also had an impact on hotel trade worldwide as Asian owners responded to the crisis by selling assets in the peaking markets of North America, Europe and Australia

This chapter concerns the impact of the 1997 Asian monetary crisis on the tourism industry of the Indonesian island of Bali. The Indonesian experience is especially noteworthy because, in addition to suffering a financial collapse, the country has undergone a political transformation. Among the countries affected by the crisis, Indonesia's experience was arguably the most acute with a contraction in national income of between 10% and 15% in 1998 (Pincus and Rizal Ramli 1998, 723). The monetary crisis and the ensuing economic collapse in Indonesia eventually led to the toppling of President Suharto's thirty two year old regime the following year, and the country rapidly became engulfed in turmoil. The impact on the country's tourism industry was apparent from the outset as visitor arrivals of 5.04 million (5,036,000) for 1997 fell short of the projected figure of 5.3 million (*Travel and Tourism Intelligence* 1998, 89; *Travel Asia* 1998). The decline accelerated as the crisis continued unabated into 1998 and 1999.

What is interesting about Bali in this context is the number of articles that appeared following the onset of the crisis stressing the island's safety. Many tourism analysts have noted that Bali has remained remarkably quiet and stable, and has continued to attract tourists (e.g. *Travel Asia* 22, May 1998), and the island's image as an exotic haven seems to have been un-dented by the strife that has engulfed the rest of Indonesia. Explanations for why this should be the case are usually couched in cultural terms, namely that Bali's population is Hindu and peace loving. If Bali really has succeeded in preserving its tourism industry at a time of great political upheaval, then this presents tourism researchers with a striking anomaly within the study of tourism. This chapter charts the impact of the Asian Crisis on Bali's tourism industry and investigates the claims concerning the island's security at a time of great political upheaval. In particular it asks whether or not cultural factors alone can account for the relative peacefulness of the Balinese as compared with other Indonesians.

There are good reasons for considering alternative explanations, not least the fact that tourism is used to manage Bali's relations with Jakarta. A theme running through Picard's exemplary work on tourism in Bali is the way this industry has enhanced the islanders' leverage over Jakarta (Picard 1996). What remains unclear, however, is precisely how this influence has been achieved, and it is the intention here to revisit Picard's analysis within the context of the Asian crisis.

In particular, this chapter argues that Bali, already something of an offshore haven for Jakarta before the crisis, has become a refuge for a variety of different constituencies since the fall of Suharto. Bali, with its good air links to the major Indonesian cities and numerous hotels, provided many Chinese with sanctuary at the height of the riots. In comparison with neighbouring Java, the island remained largely strife free and tourism and other export industries (for example handicrafts) kept the economy afloat. A closer examination of how Bali works as a haven also reveals much about how the interests of these constituencies are represented in the island.

The idea that Bali is somehow qualitatively different from the rest of Indonesia, a separate state perhaps is reinforced at airline terminals. Flights to Denpasar, for example, Bali's capital city, are often advertised as flights to 'Bali', as was still the case at Changi Airport in Singapore in 2006, whereas flights to other Indonesian cities are advertised under their respective names. Despite these illusions Bali's affairs have been inexorably linked with those of its neighbours since the island's incorporation into the Dutch East Indies and its successor state of Indonesia following the proclamation of independence in 1945. Viewed from the perspective of Jakarta, Bali was at the time but one of 26 provinces within the archipelago republic, albeit the only Indonesian province whose territory comprises a single main island (Picard 1993, 92). Anomalous though Bali is, it is still not accorded special status by the Indonesian authorities like the provinces of Jakarta, Aceh (Sumatra) and Yogyakarta (Java).

Total Crisis

In a joke circulating in Bali in 1998 the term 'total crisis' (*kristal*) was used to refer to the impact of the financial crisis upon Indonesia. The joke not only referred to the constant stream of bad news, but also was an ironic commentary on the state of affairs in Bali: the crisis did not appear to be as severe on the island as elsewhere in Indonesia. The joke utilised the Indonesian love of acronyms and summed up the crisis and its impact upon Indonesia as follows: first there was *kriseco* (ecological crisis), then *krismon* (monetary crisis), this was followed by *krispol* (political crisis) and so now we have *kristal* (total crisis). The joke may have been funny because it was a way of coping with the bad news from elsewhere in Indonesia, which included riots and many deaths, whereas in Bali life seem to go on relatively undisturbed. The news that Bali was safe (*aman*) also spread to other parts of Indonesia and was a popular conversational theme, although reasons for why the island was an exception, were usually not offered.

In the early days of the crisis the Indonesian Tourism Promotion Board followed a strategy that was similar to Thailand's. A series of measures were introduced in order to resuscitate the industry and to win back visitors discouraged by the riots and political upheavals (Henderson 1999, 300). The Board launched a 'Let's Go Indonesia' campaign, but the initiative received marginal support from the Balinese authorities that opted for a Balinese focussed approach (Hall 2000, 164–65). In 1998 the Indonesian government closed down the Indonesian Tourism Promotion Board in Jakarta, partly as a cost cutting exercise, but also as an acknowledgement of the deteriorating security situation. Bali, however, continued to function as a tourism destination and although total arrivals dipped in 1998, the industry was able to withstand the total crisis (see Table 8.1).

Table 8.1 Number of Foreign Visitors

Year	Indonesia	Bali
1996	5,034,472	1,140,988
1997	5,184,486	1,230,316
1998	4,606,416	1,187,153

Source: Bali Tourism Statistics 1999: 1.

Members of staff at Singapore Airlines were quick to point out that it was safe to visit Bali, though Indonesia itself was risky. In Bali itself a common topic of conversation was the fact that there was hardly any trouble and that the island was at peace. Precisely why this should be the case and precisely who or what was protecting the island remained unclear. A popular view, almost a folk model, shared by Balinese tourism workers, tourists and representatives of the tourism industry, was that the Balinese were less prone to violence on account of their religious outlook. This was a perspective shared by Juliet Coombe writing in *TNT UK*, a popular British magazine.

> One reason Bali remained a safe and peaceful destination is the locals' Hindu faith, which makes their outlook one of acceptance and patience (*TNT UK*, 3 August 1998, 64).

Other analysts recognised the fact that the political turbulence had spilled over to Bali, but emphasized the fact that the debates had been 'orderly' and 'confined to the campus area' (*Travel Asia*, 22 May 1998). In an article for *Telegraph Travel*, Alex Spillius provided a more detailed analysis on the situation under the industry friendly title of 'Bali: We're Safe'. He argued that, although there had been student demonstrations in Bali against corrupt politicians, the protests were peaceful and had ended when all 46 members of the legislature had agreed to resign. The fact that Bali should be treated differently from the rest of Indonesia was also mentioned in the article and an industry spokesman Lothar Pehl of the Sheraton Nusa Indah was quoted as follows:

> People should understand that there are a lot of direct flights to Bali. And consulates and embassies must differentiate between Bali and the rest of Indonesia (Spillius 1998, p. 2).

There may be some merit in the view that because of cultural reasons Bali has avoided the strife that has blighted other Indonesian islands. It is also possible that the Balinese collective memory of the bloodshed that accompanied that last change of government in 1965 has acted as a restraining force. Unlike Java, which has sprawling cities and widespread urban poverty, Bali is a small and relatively prosperous island with a limited number of ports of entry. Security is also enhanced at the village level by a network of tightly knit residential units (*banjar*) that have secular and visible responsibilities that include the regulation of interaction between village members (Hobart, Ramseyer and Leeman 1996, 88).

A widely held fear amongst a large section of the Balinese population is that tourists would not come to their island if they perceived it as being unsafe. This preoccupation is apparent in official reports, in the local press and in informal conversation. Before the elections in June, 1999, for example, which were widely expected to turn violent, there were roadside signs bearing the following slogan: 'Bali is safe: the tourists come' (*Bali aman: turis datang*). In order to test these viewpoints, however, it would be necessary to conduct the kind of attitude surveys that would possibly not be given clearance by the Indonesian authorities at the present time. All research applications have to be approved by the Indonesian Institute of Sciences (*Lembaga Ilmu Pengetahuan Indonesia*), an organisation that is wary about granting permits to study religious affairs.

Not only was the relative safety of Bali a bonus, but there was also a slump in the value of Indonesia's currency. As the Australian dollar rose against the rupiah, it was widely reported in the Australian media that a holiday in Bali was becoming a bargain, and tourists from the Antipodes took advantage of the situation despite the ongoing crisis. The Australian media reported on the delight of tourists discovering that a can of beer could be had for only 50 cents on Kuta Beach and this cheerful news was picked up by the press elsewhere. If the Australians were enjoying a cheap tipple in secure conditions, then why should not others join them?

Bali's apparent peacefulness throughout this period of transition has reinforced the view that the island is not quite like the rest of Indonesia. Despite the lack of formal recognition, it would appear that the islanders have been able to secure a greater measure of autonomy in external affairs than many other Indonesian provinces, the 'Balinese' in this case being the island's indigenous social and political elites. By emphasizing Bali's separate identity and comparative security spokesmen for the tourism industry have created what might be regarded as a virtuous and self-fulfilling circle. The island's successful tourism industry is closely linked to its peaceful image and it could be argued that it behoves both the Balinese and the Jakarta government to maintain the status quo. Explanations in the media for Bali's relative calm have been couched in cultural terms, namely that the islanders are by nature peace loving. Explanations for Bali's comparative security may cite cultural factors such as religion, but these do not have to be accepted at face value.

The popularly held view that Bali was not involved in the crisis is ambiguous when one considers that the island was and remains a stronghold of one of Indonesia's largest political party, the Indonesian Democratic Party- Struggle (*Partai Demokrasi Indonesia Perjuangan*, PDIP). In reality the Balinese are not detached from Jakarta and are very concerned about being represented in the relevant ministries, particularly tourism. When the leader of the PDIP, Megawati Sukarnoputri, was preparing to run as a presidential candidate in 1999, one of the ministers of the government of Hababie's new government declared that she was unfit for high office because she had worshipped at a Hindu temple in North Bali and that an overwhelmingly Muslim country like Indonesia should only have a Muslim president. The accusation was in any case disingenuous since Megawati is a Muslim, though her mother was a Balinese Hindu. There was uproar in Bali, particularly among the youth, and the expression 'Bali Merdeka' (Freedom for Bali) became part of the rhetoric of protest. There never was strong support for an actual split from Indonesia, but the idea that independence could be invoked has become an integral part of the struggle for expressing the right of the Balinese to remain Hindu and cropped up again in 2006 in the debate about the pornography bill. The Balinese took exception to the proposed legislation against pornography, arguing that it was too Muslim orientated and would outlaw some cherished Hindu customs, as well as threatening the freedoms of the island's increasingly successful artists.

Bali was not divorced from the political turmoil that accompanied the Asian Crisis, but it appeared to remain calm whereas neighbouring Java was convulsed by a series of severe disturbances, including ongoing student protest, grudge killings in the countryside and rising unemployment. Balinese leaders and public opinion agreed that Bali was peaceful (*aman*), which fitted in well with the touristic image of Bali that was established in the 1930s, though this was far from being the true state of affairs. In reality, there have been numerous violent incidents on the island since 1997, but the authorities have been careful not to spread the news abroad. This carefully nurtured image was, however, compromised by widespread reporting of the demonstrations in Denpasar that followed the rejection of Megawati Sukarnoputri's (the leader of the PDIP) bid for the presidency by the People's Consultative Assembly in October 1999. 'Black October' as the incident became known received widespread coverage in the Australian media causing alarm among holidaymakers as offices were torched and tyres were burned in the streets. Trees were felled bringing traffic to a standstill in Denpasar, forcing tourists who were trapped in the city to trudge many miles back to their hotels in the coastal resorts. But the following day the atmosphere lightened and Radio Australia was able to report that Balinese community groups were out cleaning the streets and were quickly bringing things back to normal. What also lifted the mood was the news that the riot had not been allegedly caused by the Balinese themselves, but by unknown perpetrators whose identities still remain unknown. The image of the peace loving Balinese remained untarnished.

As the crisis continued, the Indonesian authorities appear to be condoning a contradictory policy of regarding Bali as an integral part of the nation for internal purposes, while allowing the island to promote itself as a separate entity in international circles. The Australians, New Zealanders and Japanese, however, who comprise the bulk of Bali's inbound tourists, are relatively well informed about Indonesian affairs and are unlikely to be taken in by these policies in the long term. The emergence of a 'Boycott Bali' campaign in Australia and New Zealand in 1999 in response to Indonesia's treatment of East Timor reflected the growing awareness of consumers of the connection between Bali and Indonesia. By specifically targeting Bali in this context the campaigners were drawing attention to Bali's status as an integral part of Indonesia. Continuing unrest in Ambon, combined with the negative publicity from East Timor, was thought likely to lead to a further reduction in visitor arrivals in 2000 in Indonesia as a whole (Prideaux 1999, 287), but in fact had a limited impact.

Chinese Refugees

Businessmen critical of Indonesia's regulatory environment argue that the Suharto family operated in partnership with Chinese companies in squeezing out *pribumi* businesses (op. cit). It is widely acknowledged that the success of the Chinese is bitterly resented by the indigenous, *pribumi*, population. The Chinese are subjected to bureaucratic harassment and are obliged to adopt Indonesian sounding names, as well as being expected to carry identity cards to prove their right of abode. The riots that accompanied the downfall of President Suharto in May 1998 are reminiscent of the attacks on Chinese property that occurred towards the end of the Sukarno period in 1965 (May 1978, 135). The ethnic and religious intolerance that was unleashed during the overthrow of Suharto's regime has made investors wary (Pincus and Rizal 1998, 732).

Some analysts have argued the Indonesian crisis will be prolonged if economic hardship makes the Chinese minority an easy scapegoat (Godley 1999, 53). The disproportionate economic power enjoyed by sections of the Chinese community under Suharto is an understandable source of discontent, but the violence directed against the community as a whole during the May 1998 riots led to international condemnation. The attacks have been widely blamed on clandestine military operations manipulated by General Prabowo, the former son in law of President Suharto, but prominent Muslim intellectuals have also suggested that some kind of 'affirmative action' may be needed to redress the economic imbalance between the Chinese and *pribumi* (indigenous) populations.

The largest contingent of refugees in Bali appears to have come from Surabaya, though Chinese from other major centres such as Jakarta, Surakarta, Yogyakarta and Medan (North Sumatra) were also well represented. Initially many Chinese businessmen simply wanted to find a safe haven for their families while they continued to manage their interests in Java by making use of Bali's efficient communications and transportation. But with their property rights still vulnerable in

Javanese cities, many Chinese have decided to remain, perhaps indefinitely, on the island. Some have been attracted to the light industrial centre of Gianyar, but many have chosen to remain in South Bali and invest in the industries associated with tourism. The influx of Chinese settlers and their attendant wealth into an already densely populated and highly urbanized part of Bali seems likely to put additional strains on the environment and infrastructure. In addition to the rising demand for rented accommodation, there was an increase in the number of people wanting to purchase land, which was seen as a safe investment. The price of land rose sharply and the islanders were startled to find that a great deal of fertile land was being earmarked for yet more tourism development. This is not only occurred in hot spots such as Seminyak to the north of Kuta, but also in rural parts of Tabanan and along the coast to the temple Tanah Lot and as far as Kerambitan. The authorities opposed many of these proposed developments, but since much of the land, especially in the south of Nusa Dua and in Serangan island, had been purchased by the Jakarta-based conglomerates, the developments never took off. As the Asian Crisis unfolded investors found it increasingly difficult to access money and much of this land remains abandoned.

The Jakarta Conglomerates

The indigenous Balinese have retained a substantial stake in their tourism industry, particularly with regard to small-scale hotels, shops and restaurants. Many of the larger hotels and resort-style developments, however, have been developed with funding from Jakarta and abroad. Senior Balinese politicians such as Ida Bagus Oka, who was Governor of Bali from 1988 to 1998, also encouraged this inward investment. The influx of external capital and know-how was not necessarily welcomed with open arms and the Governor was widely criticised during his term of office for selling off the island to 'foreign' interests. While it was hard for the governor to actually oppose these investments, he did try to draw attention to the needs of the islanders by saying that the investors should develop Bali for the benefit of the community and not just practice development in Bali.

The governor was however dealing with some very powerful vested interests and there were persistent rumours that Jakarta based conglomerates associated with the family of former president Suharto had huge investments in upmarket tourism in Bali. These stories appeared to be vindicated by the publication of the results of an investigation into Suharto's assets by the American magazine *Time* (24 May, 1999), which alleged that over thirty years the Suharto family had accumulated $2.2 billion in hotel and tourism assets, though their holdings were in decline. The family had allegedly worked hard behind the scenes to safeguard their investments through the use of holding companies to blur direct patterns of ownership, making it difficult to link these assets directly to the family. The Cliff Hotel, however, the magazine claimed was owned by Sigit, Suharto's eldest son, and was one of several hotels in Bali that the Suhartos at least partly owned (Colmey and Liebhold 1999, 38).

Time's research also lent credence to the assertion made by George J. Aditjondro (1995) that five star hotels in Nusa Dua, Jimbaran and other locations had Suharto connections. These hotels were used to host national and international gatherings that had enhanced the prestige of Indonesia while simultaneously enriching the Suhartos. Not only did the Suhartos have investments in hotels, but they also were thought to own other industries linked to tourism. Sempati Airlines, for example, which was no longer in operation, seemed to have been owned by a combination of the Indonesian army and Tommy Suharto.

Aditjondro went further than other commentators in linking members of the Suharto family with specific conglomerates associated with resorts and other businesses in Bali. The list is dramatic, but difficult to verify with certainty given the reasons outlined above by *Time's* researchers (see table below).

Table 8.2 Links Between Suharto Family, Companies and Tourism in Bali

No	Name	Company	Relation to Suharto	Property
1	Siti Hardiyanti Rukmana (Tutut) b. 1949	Citra Lamtoro Gung	daughter	Nusa Dua Beach Hotel
2	Sigit Harjojudonto, b. 1951	Arseto	first son	Bali Cliff Hotel
3	Bambang Trihatmojo, b. 1953	Bimantara	second son	Sheraton Nusa Indah Resort, Bali Intercontinental Hotel
4	Hutomo Mandalaputra Suharto (Tommy), b. 1962	Humpuss	third son	Four Seasons Resort (Jimbaran)
5	Ari Haryo Wibowo	Arha	Son of Sigit/ Suharto's grandchild	Liquor imports
6	Sudwikatmono		Half brother	Nikko Royal Hotel
7	Sukamdani S. Gitosarjono			Sahid Bali Seaside Hotel
8	Probosutejo	Mercu Buana		Inter Laser Travel Bureau

Sources: Adidjondro 1995

There is some doubt, however, about the above claims since, for example, the Nusa Dua Beach Hotel was owned by a company linked to *Garuda*, the national airline, and not by Suharto's daughter, Tutut, until it was sold in 1991 to the Sultan of Brunei. One of Bali's most controversial developments, the Bali Nirwana Resort that overlooks the sacred temple of Tanah Lot, has also been linked to members of Suharto's family. *Contours* argued that the Indonesian owners of the resort, the

Figure 8.1 Dreamland Beach in South Bali near Uluwatu. Tommy Suharto owns the adjacent land

Bakrie brothers, had prospered as result of their ties to Suharto's children and his half brother, Sudwikatmono. The publication also mentioned the involvement in Tanah Lot of a former Suharto minister, Tungky Ariwibowo, a business partner in Tommy Suharto's Sentul racing circuit (*Contours* 1996, 2). The same publication also tied Tommy Suharto to the 650 hectare Pecatu Graha Project and two Four Seasons Regent hotels (ibid.). The situation appears to have been further complicated by the involvement of a British based company, Time Switch Investment Ltd, which was credited with having an 80% stake in the hotel and golfing project by Tanah Lot (*Down to Earth* 1994, 19). Both the military and the political party Golkar allegedly strongly supported the Bakries' project (ibid.).

Adidjondro also linked the Suharto family to another controversial project, the development of Turtle (Serangan) Island in South Bali. Bambang had allegedly been involved in this costly and as yet unfinished project that has angered environmentalists. By building over the island and linking it to the mainland, the developers are believed to have altered the direction of sea currents, which in turn has led to erosion and the destruction of wildlife habitats.

The conglomerates clearly had a vested interest in protecting their Balinese assets, but one should not lose sight of the fact that there was a growing substantial small and medium sized tourism sector in Bali, much of it locally owned. Luxury enclaves such as Nusa Dua only had limited links to the domestic economy of Bali, but many of the facilities in Kuta, which were used by lower budget mass tourists, particularly from Australia, were in Balinese hands (Bras and Dahles 1999, 37). These businesses were as much at risk in a downturn in tourism as the larger concerns and thus their owners had a stake in preserving the image of Bali as a peaceful destination. What should also not be overlooked is that the Balinese owned some of the larger hotels and not just the Jakarta or foreign-based investors.

In the Eye of the Storm

The fact that tourists continued to visit Bali, albeit in reduced numbers, at a time of great political turbulence in Indonesia, is a remarkable phenomenon and one that raises questions about the relationship between travel and security. This unusual state of affairs has not been subjected to a great deal of serious scrutiny, and it has largely been left to the media to provide some kind of interpretation. Nearly all the commentators opted to account for Bali's comparative peace and calm in cultural terms, namely the peace-loving and Hindu orientation of the islanders, but what appears to have escaped the attention of these authors is that Bali's history has not been especially peaceful and that there have at times been periods of great bloodshed such as suicides of 1906–1908 and the massacres of 1965 (Robinson 1995). There have also been violent outbreaks since the onset of the Asian Crisis, and though most have received little international media attention, the pro-Megawati riots of 1999 were televised widely. It is possible that other cultural factors, such as the cohesive structure of Balinese residential units (*banjar*) have helped to keep tensions in check, but this is at best a partial explanation.

Bali's comparative security at a time of great turmoil in Indonesia is less surprising when one considers that the island has become something of an offshore haven for Java. Various Jakarta conglomerates can be linked, albeit through complex business networks, to luxury hotels and other leisure enterprises in Bali and they largely remained open for business following the onset of the Asian Crisis. Furthermore, large numbers of Chinese sought sanctuary in Bali following the anti-Chinese riots that accompanied the crisis and many have stayed on to pursue their business activities in the comparative security of Bali. The fact that the Balinese themselves also held a considerable stake in the island's tourism industry doubtless also acted as an additional restraining force.

The policy of decoupling Bali from the rest of Indonesia for marketing purposes was working up until 1999, but was tarnished when a tourism boycott specifically targeted the island in the wake of the referendum on East Timor. It remains unclear, however, who was responsible for the media campaign, though some links can be drawn to the Jakarta-based conglomerates. Spokesmen for the industry stressed the special character of Bali as compared with elsewhere in Indonesia, and it is significant that one of them, who was quoted in *Travel Telegraph,* was also employed by a hotel associated with Bambang's *Bimantara* group. Support for this approach to marketing Bali was not, however, confined to the conglomerates alone and there appears to have been a much wider platform that included substantial numbers of influential islanders. Supporters included not only local owners of businesses associated with tourism and their numerous employees, but also significant numbers of what might be called opinion-leaders in other walks of life, notably politics, academia and the local press. The preoccupation with security was apparent in various interconnected ways: signs calling for restraint at roadsides, coverage in local newspapers, and informal conversations with Balinese working in hospitality. The manner in which security was managed throughout the crisis lends credence to Picard's view that so valuable was the perception of tourism in Indonesia that it could be used as a lever at the national level to secure benefits for the Balinese.

What emerges from this discussion is a picture of an island that was not as peaceful as it appears on the surface, but which was nonetheless able to manage a relatively stable tourism industry. The relative calm and peace of Bali, as compared with the rest of Indonesia, is attributable to various intersecting interlocutors: the all-important tourism, the vested interests of the Chinese and other powerful Indonesians investors. These outside interests were moreover closely interwoven into the island's economy via well-connected Balinese. These interest groups were also able to convince the media that Bali remained safe, despite the attention focused on Indonesia's troubles between 1997 and 1999. What should also not be overlooked is the powerful resonance of the name 'Bali' and its long established association with luxury and exotic travel. The prevailing view that security was vital to the success of tourism appeared to be in need of qualification, but, despite the fact that Bali remained a place with many special characteristics, when the island's security was directly threatened by bombers in 2002 and 2005, tourism went into a sharp decline. Thus, it is not the Asian Crisis and national total crisis that hit Bali's tourism hard, but rather the global dimension of the terrorist attacks.

Chapter 9

The 2002 Bali Bombings Crisis

Tourism arrivals plummeted in the aftermath of the bombings of 12 October 2002 and this had a marked impact on the island's economy and threatened to disrupt inter-communal relations. When it transpired that all the alleged bombers were Muslims, there were fears that Hindu Balinese might seek retribution by persecuting the island's minorities, but despite widespread unemployment and a collapse in living standards the island saw very little inter-communal strife. Well aware of the potential for discord, Bali's politicians and opinion leaders called for restraint using all available media, and Bali's network of village councils and urban wards. Mindful that a breakdown in inter-communal harmony would prolong the recovery of tourism, Bali's opinion leaders instituted a series of measures based on local custom and practice, and religious values, to promote inter-communal peace. To help manage the crisis, strategies such as inter-religious worship and village security capacity building were encouraged and the adoption of such measures helped to restore confidence in Bali's tourism industry. Rumours of 'ethnic cleansing' and the forced repatriation of non-Balinese, especially Muslim Javanese appear to be unfounded.

The bombings had a dramatic impact on Bali's economy, but left the Balinese 'relatively unscathed', especially those living in closest proximity to where the explosions took place (Couteau 2003, 43). The explosions happened outside the traditional area of Balinese settlement and the victims were predominantly overseas visitors; none of the Balinese fatalities included members of the local communities of Legian and Kuta. In fact, the islanders who were killed, which included members of the Muslim Balinese minority, were outsiders to the Kuta area and thus the local community did not feel that the attacks were directed against them, though they were well aware that the bombings would seriously impede the island's tourism industry as a whole. Local political leaders and the press reacted to the bombings with utmost caution, not least because there were fears before the bombings that Bali was heading towards serious inter-ethnic unrest and possibly open conflict. Bali's successful tourism industry had long attracted economic migrants from elsewhere in the Indonesian archipelago, and there had been many cases of the lynching and the scapegoating of inter-island migrants (*pendatang*), as well as on occasion the destruction of hundreds of stalls belonging to outsiders. When it transpired that all of the bombers were Muslims from the neighbouring island of Java there were well-founded fears that an economic crisis caused by the decline in tourism could ignite inter-communal conflict. Leaders of the national and local Muslim community closed ranks in their condemnation of the bombings, and the most prominent local newspaper, the *Bali Post*, long a vehicle for Bali's ethnic revival, took care to avoid apportioning blame. In addition to calls for restraint from political leaders and the media alike, there was the widespread adoption of specific cultural measures to counter the crisis avoid inter-communal conflict (Jenkins 2004).

Ethnic and Religious Diversity

Hindus account for 92% of the population of Bali, though there are small numbers of Balinese converts to Christianity and Islam. Balinese ethnicity may be defined by language and religion, but not invariably by residence since there are populations of Balinese descent elsewhere in the archipelago, most notably in the western half of the neighbouring island of Lombok, which was settled by the Balinese in the 17th century. Since independence Indonesia has also followed a policy, known as transmigration and inaugurated by the Dutch, in which settlers from the most crowded islands of Java and Bali, and to a lesser extent Lombok, have been sent to populate more sparsely populated islands such as Borneo (Kalimantan), Sumatra, Sulawesi and, while it was part of Indonesia, East Timor. There are also Balinese communities in Nusa Tenggara (islands east of Bali) where many of the settlers run concerns associated with tourism (for example hotels, restaurants, transport). The presence of Balinese in these islands may be linked to service in the Indonesian army; Balinese soldiers may be posted to the east where they eventually, bring their families from Bali to join them. Additionally, there has been a great deal of voluntary migration to the capital, Jakarta, with some Balinese taking up senior positions in the Indonesian government (the former Minister of Culture and Tourism is Balinese). Bali is itself also a destination for migrants and, as noted in Chapter 5, the island's successful tourism industry attracted a diverse workforce from across Indonesia (for example Dahles 1999; Shaw and Shaw 1999, Hitchcock 2000).

Providing an accurate account of Bali's ethnic mix is not easy since the available data are not up to date and in any case only provides information on religious affiliation. According to the census of the Province of Bali published in 2001, the population comprises 3,063,031 of which 92% are Hindu, 6% are Muslim and the rest are Buddhist and Christian (see Table 9.1). Bali's ethnic diversity cannot be described numerically, but it would be reasonable to suggest that the largest group after the Balinese themselves comprise the Javanese, followed by their close relations the Sundanese, whereas in coastal areas there are long established communities of Makassar and Bugis traders. Three predominantly Christian peoples from Flores, West Timor and North Sulawesi are also prominent in urban areas and are closely involved in the tourism sector. There are also expatriates, mostly of European and Australian origin, employed at the senior managerial level in hotels belonging to international hotel chains and the Jakarta-based conglomerates (Hitchcock 2001, Aditjondro 1995), as well as in myriad small businesses including export companies, tour companies, shops, bars and restaurants.

Table 9.1 Bali's Religious Mix

Year	Hindu	Muslim	Buddhist	Protestant	Catholic	Total Population
2001	2,823,173	183,977	18,844	21,255	15,782	3,063,031

Source: Bali in Figures, 2001

Also prominent in business are a diverse group who are not classified as indigenous, *pribumi*, though most were born in Indonesia, and who are referred to generally as the Chinese (*orang Cina*). They are mainly located in the southern urban areas and are largely the descendants of 19th and early 20th century migrants from South China. The ones who speak a Chinese dialect commonly know Hokkein and Cantonese, though the majority counts the national language, Bahasa Indonesian, as their mother tongue. The Balinese intelligentsia regards the Chinese as a well-integrated population, having held posts as trade officials in pre-colonial times (that is, before 1906) such as *syahbandar* under the Raja of Badung. Subsequent arrivals were absorbed into the existing Chinese population, inheriting a tradition of harmonious relations with the majority population, and they spread across the island as economic opportunities arose beyond the urban and coastal areas into the island's hinterland. Cordial relations between the Balinese and Chinese are often expressed in terms of a shared outlook in the spiritual world and the incorporation of selected Chinese elements (for example Chinese coins) into Balinese religious observance. The Balinese also have spaces in certain temples where Chinese Buddhists are welcome to worship. The famous Barong costume and other aspects of Balinese performing arts are possibly of Chinese origin, and there are references to the Chinese (for example *Sam Pek Eng Tae*) in Balinese legends (Spies and de Zoete 1938, 18).

The numbers of Chinese on the island were swelled during the fall of Suharto by migrants and this migration had the potential to increase inter ethnic tension due to a formidable rise in land and house values caused by the influx of Chinese capital, but these changes passed off peacefully. Intermarriage between Balinese and Chinese, although not common, does occur and there are liaisons of this kind between some prominent people in Balinese society. The Chinese are deeply involved in the island's tourism industry and are represented in all sectors: hotels, ticketing, restaurants, transportation, and clubs. It is also well known in Bali that the Sari Club, which was targeted by the bombers on 12 October, was owned and managed by a Chinese, though the employees were mostly Balinese, around ten of whom perished in the explosion.

The so-called *pendatang* or 'new arrivals' are heavily represented in the informal sector and have been a source of some anxiety among the island's intelligentsia. Cukier and Wall's survey, for example, showed that 85% of the vendors in Sanur and 73% in Kuta were non-Balinese with 68% coming from Java in the case of Sanur and 73% in Kuta (1994, 465), though the survey did not differentiate between the Javanese and Sundanese. Some linkages between employment and ethnicity have been observed (Hitchcock 2000, 218–20) with, for example, non-Balinese beach boys tending to be Javanese and Sumatran, with the latter group of mostly of Batak and Minangkabau origin. The origins of other beach boys include Sunda (West Java), Madura and Lombok. Young women working in bars could mainly be identified as Javanese, with many claiming to be from the East Javanese region, which lies close to Bali, of Banyuwangi. The people often popularly known as Madurese in Bali include significant numbers of Raas islanders who often worked closely with Chinese suppliers selling watches, sunglasses, wallets and rings (de Jonge 2000, 81–84).

Impact of the Bombings on the Economy

Since October 2002 Bali has been buffeted not only by the crisis brought about by the appalling loss of life in the bombings, but a series of other international incidents, notably SARS and the Iraq War, all of which have had a major impact on global tourism. International arrivals as measured by immigration at Ngurah Rai airport declined sharply in 2003 and were running at around less than half those of the previous two years (see Table 9.2). Visitor arrivals in May 2003 were 47,858 as compared with 119,284 in 2002 and 111,115 in 2001 and hotel occupancy, which often only ran at around 80% in the high season, rarely rose above 40% in most hotels and as low as 10% in some particularly badly hit areas such as Nusa Dua in November. By 2004 the situation had improved markedly with foreign arrivals returning – and in some cases exceeding (see January 2004) – the pre-bombing record and temporarily remained buoyant, not least because the presidential elections of 5 July and 20 September passed off smoothly with no serious incidents being recorded (*Bali Rebound,* July 10 – July 24).

Despite the upturn, such was the severity of the crisis that most hotels and tourism related businesses had to either lay off workers or re-employ them part-time. Grand Mirage Hotel in Tanjung Benoa (Nusa Dua), Sobek Rafting (Sanur), Water Bom Park (Kuta), Sahid Bali Hotel (Kuta) were among those who had to shed large numbers of staff. The management of Sahid, for example, cut around a hundred jobs saying that they were surplus to requirement and that they would not re-call them. Staff cuts at Sobek were achieved through 'golden handshakes', in which staff members were voluntarily – and in some cases involuntarily – retired. Retiring employees were also rewarded with bonuses and because these redundancies were technically voluntary, they were not obliged to report to the Department of Employment (*Dinas Tenaga Kerja*). They were therefore not officially recorded as having lost their jobs due to the crisis and such job losses were not regarded by the labour unions as being an infringement of employment laws.

As a result of the crisis some properties in Kuta were placed on the market, but such transactions are hard to verify because of business discretion. Despite the lack of data, the sale of property remains a prominent issue among local business people. In East Bali, particularly in the Candidasa beach area, one of the popular tourist destinations with middle and low budget hotels, many hotels and bungalows were put up for sale (*Bali Post* 2003). These sales were a direct result of the bomb blasts and other crises, leading to occupancy rates of 20%, when the breakeven point for many is an occupancy rate of 40%. Nyoman Ruta Adhy, The Head of the Hotel and

Table 9.2 Direct Tourist Arrivals Before and After the 2002 Bombings

Direct Foreign Tourist Arrivals to Bali by Month								
	1999	2000	2001	2002	2003	2004	2005	2006
January	102,280	92,604	108,897	87,027	60,836	104,062	101,931	79,721
February	105,240	104,083	99,040	96,267	67,469	84,372	100,638	73,430
March	117,172	110,582	115,997	113,553	72,263	99,826	117,149	84,109
April	104,028	109,634	117,040	104,960	53,726	111,022	116,272	103,886
May	104,526	103,939	111,115	119,284	47,858	117,191	116,615	101,776
June	119,357	122,352	128,792	130,563	81,256	131,707	136,369	109,651
July	143,920	142,946	138,150	147,033	111,828	148,117	158,453	121,988
August	146,209	144,324	145,290	160,420	115,546	155,628	157,229	118,104
September	134,688	104,008	133,667	150,747	106,763	141,952	162,102	118,331
October	104,251	129,932	96,537	81,100	97,435	128,399	81,109	112,629
November	87,763	110,145	72,806	31,497	83,853	110,506	62,705	113,844
December	86,365	102,290	89,443	63,393	95,783	125,525	75,877	122,848
Total	1,355,799	1,376,839	1,356,774	1,285,844	994,616	1,458,309	1,386,449	1,260,317

Sources: Bali Tourism Authority (2002), *Bali Travel News* (May 26–June 8, 2006) and Bali Tourist Board 2006

Restaurant Association of the Badung Regency, where most tourism activity takes place, is aware of one hotel closing down in Kuta and the sale of many more. Most hotels were partly closed in 2003 as their managers tried to cut overheads on water and fuel by switching off their air conditioning.

Solid data on salaries is difficult to obtain in Indonesia, which is partly a legacy of the secrecy of Suharto's regime and partly because of the lack of efficient statistical recording, but it would appear that when people are lucky enough to have a job then they were only being paid the basic salary, which is often the legal minimum wage. Workers are usually rewarded with extra service fees and bonuses and thus their remuneration in the 2003 high season may represent a drop of between 40 and 50% as compared with their normal pre-crisis salary.

The multiplier effect by which the benefits derived from tourism are spread through the economy in general works in reverse in a crisis, and generally economic activities were slow or much reduced following the onset of the crisis and affecting not only people employed directly in tourism, but also those who work in related sectors. Taxi drivers reported finding passengers hard to find and restaurants and souvenir shops had few customers. Numerous shops in the main resorts of Kuta and Sanur were either closed down or mothballed as their owners waited for good times to return. Even businesses with no clear linkages to tourism reported a downturn. Garages, for example, experienced declining demand since fewer people had sufficient money to pay for oil changes and car services.

Business activities in the tourism sector had started to recover a little by mid-2003, but the rate of increase was still well below the level of the pre-bombing era. Increasing confidence among tourists to visit and stay in Kuta has led to the re-opening of many bars and restaurants, which had been closed since the bombing. Among those re-opening was Paddy's Bar, which had been made famous because of the bombing, in a new location barely a hundred meters from the original site (Chulov 2003). The new location of Paddy's is situated on the same spot as the Bounty Ship, which is owned by Kadek Wiranatha. He is also the owner of the airline, Air Paradise International Bali, which went into service on the Bali-Australia route on 16 February 2003. The new service had been delayed from its initial start date of 24 October 2002 because of the explosions. With the opening of New Paddy's almost all of the 120 staff of the old Paddy's, none of whom was killed by the blast, returned to work. In contrast to the Sari Club workers who remained unemployed, they were lucky to get their jobs back after suffering more than eight months redundancy. Returning employees in other hotels and restaurants found that their wages were much diminished because the businesses they worked for were not as busy as before and consequently received less income from service charges.

Regional and National Discourse

As the identities of the bombers were discovered and it emerged that all were Muslims, there were fears among Indonesian officials that this information would ferment inter religious unrest in the manner of the Indonesian island Ambon, which had experienced severe Muslim and Christian clashes since the late 1990s. The possibility of an anti-Muslim backlash also seems to have stiffened the resolve of Bali's opinion leaders and provincial officials to avoid inter-religious and inter-ethnic conflict at all costs. Bali's and Indonesia's leaders were well aware that even small clashes could lead to a long term deterioration in relations that would not only prolong the time taken for tourism to recover, but might kill off the industry all together. The measures adopted to avoid conflict were similar to those used during the Asian Crisis when signs were posted throughout the island that said '*Bali aman, turis datang*' (Bali is safe, the tourists come). On the instructions of the national government, banners were posted prominently, mainly in Kuta and Sanur, with the slogan 'Bali for the world', a slogan that was widely rejected in Bali; it was seen as too Jakarta oriented and thus out of touch with local desires and perspectives. One of the reasons for the ambivalence on the part of Balinese stakeholders in tourism was that the recovery programme appeared to have been conceived largely from the perspective of potential beneficiaries in Jakarta. Many interpreted the slogan as meaning that Bali was 'for the world' and not for the Balinese and thus disliked it (Aridus 2002; Wedakarna 2002), and this was coupled with an emerging sentiment that local stakeholders should have more control over the island's destiny and of any recovery programme.

An underlying unease about the implementation of regional autonomy in contemporary Indonesia can be detected in this regional-national discourse because Bali's borders have, unusually for an Indonesian province, matched those of the island. There were widespread fears that devolution to the regency level could disrupt Bali's coherence, which was not only a cultural and ethnic issue, but is also an important consideration for the effective management of tourism, since many leisure activities cross regency boundaries. The intelligentsia widely believed that Bali's unitary authority had the potential to serve the island well both culturally and strategically in terms of the island's recovery and future development.

A related concern was the rise of the Bali independence movement, *Bali Merdeka* (Freedom for Bali), which had a minor mandate prior to the crisis. The rise in the independence movement was fueled by criticism in mid 1999 by one of former President Habibie's ministers of Megawati, who subsequently became president, for praying at her grandmother's Hindu temple in North Bali. Megawati who is a Muslim of part Balinese descent was accused of being unfaithful to Islam and thus incapable of serving overwhelmingly Muslim Indonesia as President. This criticism caused widespread offence in Bali, but the overwhelming majority of islanders realize, that an independent Bali is not realistically feasible; the aims of *Bali Merdeka* have become a form of rhetorical weaponry to be used against Jakarta, though the independence movement has waned (Vickers 2000). In the aftermath of the crisis it was not the independence movement that articulated Bali's sorrows and frustrations, though its message simmered away, but the increasingly outspoken newspapers and other media on the island.

Security and the Street Traders

The tourism authorities have long been apprehensive about the activities of the unlicensed hawkers and vendors of the informal sector, not least because of complaints about the aggressive behaviour of street and beach traders. Dissatisfaction with their activities had featured highly in surveys of visitor satisfaction and had been reported in both the local and national press, and – more worryingly – abroad. The situation was, however, not clear-cut and tourism officials readily admitted that there were small numbers of visitors, some of whom were undoubtedly lonely, who liked interacting with street traders, though the majority clearly disliked them. A prevailing view among government officials was that the anarchic behaviour of pushy traders made many visitors uneasy and potentially heightened their concerns about security. They were also aware that many of these traders were from elsewhere in Indonesia, but tourism officials repeatedly stressed that non-Balinese Indonesians were welcome, and indeed had the right as national citizens to be in Bali, providing that they traded legitimately with the correct permits. For a long time the tourism officials had wished to 'clean up' the island's business activities by issuing licenses and enforcing trading regulations. Despite this stance, however, there had been rumours that traders in the informal sector have been able to pursue their illegal activities by bribing the very police officers whose job it was to close them down. Such is the ubiquity of small-scale corruption in Indonesia that it was not possible to corroborate these rumours with any certainty for fear of singling out individual policemen and thereby undermining attempts by the authorities to upgrade the island's security. In post-Suharto and reformist Indonesia it was at least possible to acknowledge that ongoing levels of petty corruption were encountered in tourism related businesses.

Security has been given a top priority in the wake of the bombings and anything that detracts from the desired image of Bali as a safe haven is viewed with circumspection. Inspector General I Made Mangku Pastika, the police chief who led the investigation, was given a clear mandate from the island's governor to do all in his power to look after the safety of visitors. Pastika has presided over a complete change in policing policy on the island in response to requests from businesses and the public alike, but he has inherited a legacy of petty corruption and police inefficiency dating from the Suharto era. The management of the post crisis recovery also raised concerns about the image of the Indonesian police force since before the bombings many hotels and shopping centres were unwilling to see uniformed police protecting their premises because they felt that it heightened the sense of insecurity. Prior to the crisis the emphasis was on covert policing and responding to crimes by calling on policemen as and when they were needed. The police were often out of sight and confined to their stations and barracks waiting to attend emergencies as they arose. As the recovery continued, visible policing became a priority and this also spurred debate about issues such as the appearance of men in uniform.

The tourism authorities had long wanted more user-friendly uniforms and even specific tourism training for police officers and used the crisis to press home their views with some success. The first immediate outcome was the tripling of the numbers of intelligence officers, partly because of renewed fears about a fall-out from unrest in Aceh and increasing concerns about the possibility of further terrorist attacks. The additional officers were posted at all points of entry, notably the international airport of Ngurah Rai and the harbour of Gilimanuk, the latter being the main entry port from Java, the route taken by Amrozi and his fellow bombers. The second outcome was the inauguration of beach policing, *polisi pariwisata* (tourist police), on Kuta Beach in the 'Baywatch' style associated with the USA and Australia. Foreign expatriates volunteered to upgrade the use of English of the new Baywatch squad and to teach them Western etiquette. In the short term an accreditation scheme of security standards was also introduced to ensure that police were aware of the layout of hotels in the event of an emergency. Advisors from Japan and Australia were also brought in to help upgrade the island's security. The new security measure eventually led to a system of accreditation for hotels run by the police, which included improved mapping of hotels and resorts.

Among the first to bear the brunt of these changes of emphasis in policy were the street traders and instead of waiting to be registered, many chose to leave the island, presumably to return whence they came. The combination of the downturn in tourism and the fact that licensed traders are liable to taxation were doubtless other factors that led to the outflow. In contrast the Balinese masseurs who ply their trade on the island's beaches have long been officially licensed and are obliged to submit tax returns, albeit somewhat erratically; to promote easy identification many have numbers painted on their straw hats. There was undoubtedly a mass exodus of petty traders following the onset of the crisis, accompanied by grisly rumours of truckloads of them being rounded up by the authorities, but there appears to be no hard evidence of enforced repatriation. The combination of General Pastika's crackdown on licensing and the rapid decline in the numbers of visitors simply persuaded large number of traders in the informal sector that it was no longer worth staying in Bali and it remains to be seen, if and when tourism picks up, whether they will return.

Cultural Responses to the Crisis

In addition to the largely practical measures discussed above, the Balinese adopted a number of local measures to combat the crisis and spur the recovery, albeit controversially in some cases. Various measures were inaugurated to promote inter-religious harmony, most notably a series of inter faith dialogues, which included joint prayers. These started with a prayer for world peace, *Doa Perdamaian Dunia dari Bali*, on the afternoon of the auspicious full moon of 21 October, 2002, which was attended by the Minister of Religious Affairs, who was a Muslim. Inter-faith prayers were held at various locations including the bombsite at Kuta Beach, at both football fields in Denpasar and in the most holy temple in Ubud. These observances have continued with an annual inter faith commemoration at the Denpasar football fields.

One of the spiritual measures that was undertaken was televised abroad to rapt audiences around the world were the elaborate purification rituals, usually involving huge processions of traditionally clad islanders bearing offerings.

Despite being publicized widely in the international media, there was often little interpretation and it remains unclear what audiences from many different cultures made of these festivals. What was often overlooked was that the islanders had ended up construing the bombings as an expression of the anger of the gods, a consequence of the bad karma suffered by Bali for having '… allowed impurity to hold sway over the land' (Couteau 2003, 45). In order to resolve this inauspicious condition a huge ceremony known as the *Pamarisudha Karipubhaya* was held on 15 November 2002, involving the most powerful priests, and using the most efficacious mantras and holy waters that the island could provide. For adherents of Balinese Hinduism these huge demonstrations helped to cleanse the trauma of the Kuta bombings, and as some local leaders opined, enabled the Balinese to put these terrible events behind them and to contribute effectively to the emerging recovery strategy. The focus on impurity and ritual helped to reduce the apportionment of blame to other religious groups and provided a cathartic outlet for potentially negative energies.

Bali's performing arts are not solely entertainment and many have ritual connotations, and so they too were utilized in the recovery process. For example, it is customary for the comic figures in Bali's masked dance-dramas and the shadow theatre, which celebrate the ancient Hindu epics, to provide humorous commentaries on contemporary affairs. References to the bombings, tourism and the need for harmony thus popped up in numerous shows in the aftermath of the Kuta tragedy. The clowns also were also involved in promoting Balinese tradition, notably values and religion, and especially the law of karma. Two authors (Jenkins and Nyoman Catra 2004, 3) have also discussed the use of Balinese culture as a weapon to counter the crisis, a strategy that has its origins in the first century AD according to local performing artists. It may be impossible to verify the antiquity of such claims, but it is clear that such cultural strategies are a long established local resource.

Another outcome of the crisis was the bolstering of various Bali's traditional institutions, notably the network of village sub groups or wards, known as *banjar*. It was widely agreed that the *banjar*s, with their grassroots membership, were an indispensable weapon in the fight against terrorism, though some foreign commentators (for example Couteau 2003) have not been enthusiastic about the growing influence of the village security services, the *pecalang*. They are often untrained in security work and have been accused of resorting to thuggish behaviour, some of which has been directed towards migrants, *pendatang*, from elsewhere in Indonesia. Balinese opinion formers, however, ignore this kind of criticism of village level security by drawing attention to the success of Bali's village sub groups in helping to uncover the evidence that brought the bombers to justice and their role in restraining the islanders from seeking retribution for the bombings. In addition, there were worries that Islamic radicals would disrupt the trials of the bombing suspects and the police, fearing they might be overstretched, sought assistance from the *pecalang*, and undertook capacity building measures to boost their competence in security. Such was the growing confidence of the regular police in the *pecalang* that these auxiliaries were deployed during the visit of President George W. Bush.

Post Crisis Religious and Ethnic Diversity

The departure of large numbers of petty traders has altered the ethnic mix of the island, but since nobody was ever able ascertain precisely how many people were involved in the informal sector, the change is difficult to quantify. In contrast, the ethnic composition of the formal part of the economy appears to have remained relatively stable with large numbers of migrants from outside Bali continuing to work in licensed trades. By 2003 migrants from Raas, for example, could only rarely be seen peddling their wares on the beaches at Sanur, but legitimate Raas islanders could still be found working in shops. Likewise, many of the Bataks, Sasaks and Javanese who worked informally seemed to have vanished, though workers from these communities, albeit in reduced numbers were still playing a vital role in the formal sector. The migrants who stayed were often quick to point out that the island had become a multicultural society and that they respected the cultural values of the Balinese and were made to feel welcome.

The Indonesian economy had hardly recovered from the downturn of Asian Crisis when the bombers struck and fueled the number of requests for visas for Australia, particularly from the Chinese, many of whom had fled to Bali from elsewhere. Despite the downturn, many of the newer migrants appeared to be well on their way to becoming a permanent part of the island's business community. Many Chinese continued to invest substantially in the island's future, often repatriating investments that had flowed out of Indonesia after the fall of Suharto. This reverse flow of money owes much to improvements in Indonesia's fiscal policies since the late 1990s and some of this Chinese money seems to have been directed to investment in tourism in spite of the crisis. Members of this newly settled Chinese population are relatively open about why they chose to remain on the island, citing good relations with the Balinese as an important factor. In a similar vein, the long established Chinese stressed their history of harmonious links with the majority population in Bali and most showed little inclination to move elsewhere. The picture may be rosy when it comes to Balinese-Chinese relations, but Balinese opinion leaders do sometimes worry about the long-term impact of a substantially increased Chinese population, with some even mentioning the possibility of Bali becoming more like Singapore. Members of the intelligentsia worry about Bali losing its distinctive cultural identity and they stress the importance of differentiating the island in the marketplace, not wishing to play second fiddle to Singapore.

In tandem with the bombings crisis, there was an intensification of the emerging debate on precisely what it meant to be Balinese at a time of increasing regional autonomy within Indonesia. Some opinion formers were quite enthusiastic about Bali's multiculturalism, pointing out in particular the value of the expatriates to the economy, many of whom were clearly 'Bali lovers'. Some of the contradictions surrounding Balinese identity were expressed in a popular song created by Guruh Soekarnoputra, which was popular in the late 1980s and early 1990s (Darma Putra, 2003), and regained its popularity after the bombings and was played almost continuously on radio and television. Entitled 'Bring Back my Bali to Me' (*Kembalikan Baliku Padaku*), the song encapsulated a growing awareness of being a Hindu-Balinese minority within the Indonesian nation state. McDonalds used the song for the backing track for a commercial on national and local television in Bali, cleverly tapping into the local mood while appearing to be socially conscious. Following the bombings there was widespread support for imposing further restrictions on inter island migration, especially from Java via the harbour of Gilimanuk. Directly after the explosions restrictions were implemented quite strictly and only those with valid identity cards and guaranteed jobs were permitted to enter Bali. Hundreds of would-be migrants were turned back at Gilimanuk, but by mid 2003 this policy was no longer rigorously being enforced, suggesting that the prevention of internal migration to Bali is impossible. Despite their growing anxiety, the Balinese had no choice but to accept migrants as a reality. Whether they liked it or not the Balinese had to accept the idea that the island would in all probability become more multicultural, leading to an intensification of expression and interest in what it meant to be Balinese. The resurgent popularity of 'Bring Back my Bali to Me' represented on one hand an intensification of the identity debate with a growing realization and acceptance on the other that the island was no longer for the Balinese alone.

The Return of Security

The Balinese population as a whole appears to have accepted the principle that peaceful conditions were a prerequisite for the successful maintenance of tourism. The 2002 crisis not only appeared to have stiffened the resolve of opinion leaders and the population as a whole, but also mobilized traditional communities and leaders. The crisis was managed by combining a number of practical security measures and with some distinctively Balinese cultural strategies designed to stave off inter-communal strife. Insults were traded among hotheaded youths belonging to different ethnic groups, but there were no verifiable reports of reprisals against migrant workers, *pendatang*, though rumours of unpopular migrants being frog-marched off the island persisted. But, there were no incidents of the kind reported in 1999, when a series of lynchings took place and one death, as well as the destruction of stalls belonging to non-Balinese traders (Couteau 2003:55). The tales of mass deportations of *pendatang* were probably just rumours and what seems to have happened is that many traders simply returned home during their annul holidays of *Lebaran* (the Muslim festival of *Idul Fitri*) and never returned due to the worsening economic crisis. The decline in the number of non-Balinese seems to have largely been restricted to the informal sector, the part of the economy most vulnerable to both the downturn and the imposition of tougher trading regulations, and seems to have been far less change in the ethnic composition of the formal sector. In fact, one migrant group, the Indonesian Chinese, seemed to have forged a firm alliance with the Balinese during the crisis.

The rise in Balinese separatist sentiment in the wake of the bombings was also a reaction to growing concerns about the island's ethnic diversity, but the calls for separation from the Indonesian state were taken very seriously. There was some confusion, however, among many Balinese, as well as among some expatriates, that the growing assertiveness of the island's leaders and stakeholders in its tourism industry was an expression of separatist sentiment. What had really happened is that opinion formers in their power struggles with Jakarta had adopted some of the separatist movement's rhetoric. The separatists' slogans but not their policies had been adopted and in any case *Bali Merdeka* had very little influence since the majority of islanders continued to support mainstream Indonesian political parties.

Following the crisis, the Balinese and Indonesian opinion leaders and media commentators closed ranks to stave off calls for retribution, but it was not due to their efforts alone since their message appears to have been readily understood by the population at large. The willingness of ordinary Balinese to fall in behind local and national opinion leaders is attributable to three significant factors. First, Bali's religious leaders placed emphasis on non-violence and a respect for peoples of other faiths, and outlook endorsed by the Indonesian constitution in the state national code of *Pancasila*. As loyal Indonesians belonging to a small religious minority, the Balinese have a considerable interest in seeing that state code being maintained and are thus honour bound to live by its principles. Second, such was the importance of tourism in Bali's economy that ordinary people were highly motivated to protect it. Third, Bali had become an increasingly media aware society with high levels of access to the national and international stations, and thus the population well versed in the consequences of an escalation of inter ethnic violence in cases such as Ambon and Sampit (West Kalimantan). By and large the Balinese were proud that an incident as serious as the Bali bombings did not damage their reputation for being decent and civilized people, and that the island did not descend into inter-ethnic violence, and were keen to maintain this discipline during the run up to the legislative and presidential elections in 2004. The threat of social unrest was far less after the 2005 bombings, not only because police acted quickly to control the situation and to search for the perpetrators, but also because of the heightened awareness of the islanders of the link between security and tourism.

The sun will rise, the sun will fall. We will remember all those lost and know the terrorists will never win.

BALI AUSTRALIA

FRIENDS FOREVER

Peace and happiness to all Balinese people, and to those who come to visit this beautiful country

Love from

Gloria, John & Jordie

(Australia)

200

Figure 9.1 **Message among the wreaths outside the bombed Sari Nightclub in 2003 refers to the Australian graduation song 'Friends for Ever'**

Figure 9.2 Wreaths on the undamaged shrine above the spring in the Sari Nightclub in 2003

Chapter 10

Global Conflict and the Bombers

The tourism industry worldwide has suffered a major setback since the World Trade Centre attacks in the USA, but the impact on Asian tourism was initially moderated by the rise in intra-regional travel as Asian tourists opted for shorter-haul and destinations that were perceived to be safe. Because of the high levels of growth in mainland China, travel also became more affordable for the Chinese (*Asian Market Research News*, 30 April, 2003). The SARS crisis, however, followed by concerns on the effects of the war in Iraq and terrorism in general has affected one of Southeast Asia's major incoming tourist markets, namely East Asia including Japan, China, Taiwan and South Korea. Tourism was also shaken in the region when a militant Islamic group from the Philippines, Abu Sayyaf, took 21 hostages, including 10 foreign tourists, from a diving resort in the Malaysia state of Sabah, with the kidnap earning terrorists US$ 20 million, reportedly paid by Libya (Rabasa 2003, 54).

Following 11 September 2001 the security environment also changed both globally and regionally, as extensive terrorist networks were uncovered in the ASEAN region and countries such as Indonesia were becoming battlegrounds in the US-led war on terror. The strife is not limited to Indonesia and there have for example been major problems in Thailand and the Philippines, as well as serious alerts in Malaysia and Singapore, but what is significant in Indonesia is that one of the targets that has been consistently sought by terrorists are tourists.

Tourists are easily attacked and some of the things that they engage in may be used as a justification for attacking them and since many of them in the case of Bali come from Australia, an ally of the USA, then the terrorists are able to justify their attacks on innocent civilians in terms of a striking back at the USA. Bali is moreover an especially tempting target since it is a very well known tourism destination and any outrage that takes place there is likely to receive extensive media coverage worldwide. As a result of the explosions of 2002 at least 202 people lost their lives (see Table 10.1), though the full extent of the casualties may never be known for sure because of the difficulties in identifying all the victims. It represents not only the largest act of terrorism in Indonesian history, but also one of the largest attacks on a tourist resort.

Tourism's Sensitivity to Crises

It is widely held by tourism promotion boards, that tourism is sensitive to crises and that the press in particular has a role to play in helping alleviate the fears of travellers. Thus the media is seen as being a major force in the creation of images of safety and political stability in destination regions (Hall and O'Sullivan 1996, 107). It is not just direct threats to tourism such as terrorism that are seen as a problem, but negative reporting in general about a given destination. Thailand, for example, became so alarmed about the future of its tourism industry due to the negative press coverage of the Asian Crisis, which it was at the heart of, that it sought to counter the negative image by positively promoting the country as a cost effective destination (Higham 2000, 133). Thailand successfully boosted visitor arrivals, as a means of supporting its economic recovery, and the country has remained very sensitive about its image as a tourism destination.

Not all the strife that has a negative impact of tourism is directly concerned with tourism, though tourists have become targets to advance certain religious and political causes since the early 1990s at least, the most publicised case being that of Luxor in 1997 which left 58 foreign visitors dead. But, even before Luxor terrorists were targeting tourists and, according to the Ministry of the Interior, had killed 13 of them, as well as 125 members of the Egyptian security forces, since 1991 (Aziz 1995, 91). As has also been pointed out by Heba Aziz, Islam is not against tourism per se and she has argued that the attacks in Egypt can be understood as a reaction to insensitive kinds of tourism development and that the tourism industry, the Egyptian government, the developers and the tourists '...are as responsible for this undesirable situation as the Muslim groups' (Aziz 1995, 91). There are some elements of this in the case of the Bali bombings, but the attacks in Indonesia can also be linked to a wider trend of global terrorism that embraces a number of continents and causes, though the motives of the bombers in this case appear to have some specifically Southeast Asian causes.

The 2002 Bombings

Three targets were bombed on the night of 12 October 2002: the Sari nightclub and Paddy's Bar in Kuta and the American Consulate in Denpasar, not far from the former Australian Consulate office. The consulate bomb had no victims, but the ones in Kuta were devastating. The bomb at Paddy's Bar did not at first appear to have a great impact, but it had a deadly side effect. It drew people on to the streets so that when the next bomb at the nearby Sari Club went off more people were vulnerable. The explosives had been packed into a van that had been parked outside the packed nightclub, which was almost entirely destroyed by the blast and fierce fire that followed.

The victims were drawn from 22 nations, but the brunt of the tragedy was borne by Australia, with 88 dead. The second largest loss of life, with 35 dead, was suffered by Indonesia, the majority being Balinese islanders. What is often overlooked in reportage of the bombings was that many of the Balinese dead were Muslims, a minority in the largely Hindu island of Bali. The third largest toll was experienced by the UK, which lost 23 of its citizens in the explosion. Not only are Americans (7) and many European countries (Germany, Sweden, France, Denmark, Switzerland, Greece, Portugal, Italy and Poland) represented in the list of victims, but so are Canada, South Africa and New Zealand. There were also other Asian victims (Taiwanese, Japanese and South Korean), as well as South Americans (Brazilians and Ecuadoreans) (see table 10.1).

Table 10.1 The Victims' Countries of Origin in the 2002 Bali Bombings

No	Country	Number of victims
1.	Australia	88
2.	Indonesia	35
3.	UK	23
4.	USA	7
5.	Germany	6
6.	Sweden	5
7.	Switzerland	3
8.	The Netherlands	4
9.	France	4
10.	Denmark	3
11.	New Zealand	2
12.	Brazil	2
13.	Canada	2
14.	South Africa	2
15.	Japan	2
16.	Korea	2
17.	Italy	1
18.	Portugal	1
19.	Poland	1
20.	Greece	1
21.	Ecuador	1
22.	Taiwan	1
TOTAL		196
Plus 3 unidentified victims and 2 bombers		Grand total 201

Source: Planning Bureau of Badung Regency, 2003

At first, the Australian Federal Police maintained the attacks were the result of a high degree of planning and expertise and were designed to cause the maximum number of casualties. A spokesman on behalf of the police, Graham Ashton said the bombs were situated strategically to take advantage of surrounding buildings and that technical experts had appraised the bomb-making skills as "above average" standard (*CNN*, 1 November, 2002). The investigators believed that the larger explosion, which destroyed the Sari Club, was caused by the chlorate that had been set off by a "booster charge" such as TNT. Ashton reported that 400 kilograms (880 pounds) of chlorate was stolen in September from a location on Indonesia's main island of Java, but he declined to elaborate on the theft. The trial revealed that Amrozi also bought one ton of calcium chlorate (KClo3) in a shop in East Java, which he dispatched to Bali by bus along with other chemicals.

Figure 10.1 London's Bali Memorial, unveiled by Their Royal Highnesses The Prince of Wales and The Duchess of Cornwall on 12th October 2006

Figure 10.2 London's permanent memorial to the 202 victims of the Bali bombings

Links to Global Terrorism

In response to the large numbers of Australians killed, the Australian Federal Police joined forces with the Indonesian investigators, who were led by the able Inspector General I Made Mangku Pastika. The bombers had taken the trouble to cover their tracks, but despite their care an essential part of their plan came unstuck and this led Pastika reasonably swiftly to the first of the suspects, Amrozi bin H. Nurhasyim. He was a mechanic and had hoped to foil detection by changing the registration number on the van that he bought to transport the larger bomb. Amrozi was unaware, however, that the van had previously been used as a minibus and thus bore another number, which Pastika's investigators traced back to Amrozi. Dubbed the 'smiling bomber' by the media, Amrozi, quickly gave the police a detailed confession, after an interview with Indonesian police chief Da'i Bachtiar, which underpinned much of the investigation that followed.

Once the investigators became aware that the bombers were radical Muslims, attention was focused on their motives and potential linkages with known terror groupings. Defence Minister Matori Abdul Djalil and other Indonesian officials, who had previously questioned the involvement of al-Qaeda cells in Indonesia with the support of local activists, eventually agreed that the attacks were linked to al-Qaeda (Rabasa 2002, 34). Furthermore, intelligence sources in Indonesia identified a Yemeni al-Qaeda operative, Syahfullah and a Malaysian called Zubair as having been involved in the bombings. Indonesian intelligence believed that Zubair had been involved in surveying and mapping the targeted sites, whereas Syahfullah was thought to have helped Imam Sumudra and Ali Gufron (alias Mukhlas) to coordinate the attacks (ibid.). Through Mukhlas a connection was made to an organisation known as *Jemaah Islamiyah* (JI), one of whose main objectives was to redraw the modern borders of Southeast Asia to create a substantial Muslim Caliphate, a position steadfastly opposed by the governments of the ASEAN region.

The Australian and U.S. governments suspected that the bombings were the work of an al-Qaeda-linked terror group based in Southeast Asia, Jemaah Islamiyah (JI), but nobody at this early stage had claimed responsibility for the Bali attacks. Suspicion fell on JI's alleged leader, Abu Bakar Ba'asyir, who was arrested on October 18 after Indonesian investigators returned from questioning al-Qaeda operative Omar Al-Faruq, who had been handed over to the United States in June earlier that year. Al-Faruq claimed to know Ba'asyir well and said the cleric was involved in attacks on Christian churches in Indonesia in late 2000. Rabasa has argued that Jemaah Islamiyah was linked to al-Qaeda by Hambali, an experienced operative who faught against the Soviets in Afghanistan and was one of the few non-Arabs to have been allowed to join al al-Qaeda's supreme council, *shura* (2003, 63). The authorities in the United States, Australia, Singapore and the Philippines claimed to have evidence of JI links to al-Qaeda and that the group has established several cells throughout Southeast Asia and Australia, but whether Ba'asyir was involved in the Bali bombings remains controversial and according to a International Crisis Group report of December 2005 he is said to have opposed the bombings and was unlikely to have been the mastermind behind the attacks (Rabasa 2003, 35).

Motives and Justifications

While the shadowy links between Jemaah Islamiyah, al-Qaeda and the Bali bombers may not be fully established with any certainty, the trials of the 2002 bombers do provide an opportunity to examine the motives of the bombers with some clarity. The defendants have been interviewed by professional journalists, have written confessions, including a book in one case, and have been tried in public. What is significant is that the bombers initially claimed that they were attacking Americans, though the largest number of victims turned out to be Australians and the second largest, Indonesian. Their claim would, however, appear to be supported by the fact that they also targeted the American Consulate that night, placing a bomb about a hundred metres away from it, serving a warning to the USA that it was the target. As the casualty figures mounted and it became clear that only 7 out of the 201 victims were American, a range of other political justifications began to be offered, though they were provided with the benefit of hindsight. In particular there was an alleged statement by Osama bin Laden who said that it was indeed Australians who were being targeted because of their alliance with the USA. It would appear that the bombers had expected the club to have more Americans, but, when informed that the majority of their victims were Australian, one of them quipped: 'Australians, Americans whatever – they are all white people.'

The bombers showed no remorse and expressed their satisfaction with the number of foreigners killed, seemingly unconcerned about the deaths of the Indonesians, many of whom were Muslims. Amrozi offered to pray for the dead Balinese, but seemed unmoved in his conviction that he had done something satisfying; his only regret was that he did not kill more Americans. Amrozi and his fellow conspirators decided to set the bombs in Kuta because of the large number of foreigners there and when he heard that many of them had died he said he felt very proud. Westerners were targeted because they were perceived as being associated with attacks on Muslims, and Amrozi made clear that he felt no remorse about this.

> How can I feel sorry? I am very happy, because they attack Muslims and are inhumane (*Asia Times* 3 June 2003).

As the trials continued reporters from around the world were horrified by the trite and cruel behaviour of the bombers, with one paper likening them to the Nazi Albert Eichmann because of his complete lack of remorse for his crimes against the Jews.

While exact parallels may not be drawn, broad similarities appeared in the Bali bombing trials. The cavalier, almost frivolous, attitude toward human lives seemed to be rooted in the banal worldview of the alleged Bali perpetrators (*Asia Times* 3 June 2003).

Amrozi was quite open during the trial about what motivated him to carry out the attacks and he claimed that he had heard about the decadent behaviour of white people in Kuta from Australians, notably from his boss while he was working in Malaysia. Amrozi had worked alongside French and Australian expatriates in a quarry and this is where he learned about explosives. Amrozi also said that it was his colleagues who told him what an easy target Bali was and he became incensed by their stories of drug taking and womanising. Amrozi came to hate Westerners, believing in the conspiracy theory that the Jews bankrolled them and wanted to control Indonesia. He convinced himself that the only way to persuade Westerners to withdraw from Indonesia was through violence and rejected diplomatic means as unworkable. According to Amrozi, the bombers comprised a core group of nine who were experienced in carrying out bombings and who were united in their hatred of Westerners. Amrozi claimed to have been involved in bombings in Jakarta, the Indonesian capital, and in the fractious regions of Ambon and Mojokerto in central Java where 19 died in the Christmas Eve attack of 2000. He also confessed that he had played a part in the attack on the Philippines Embassy in Jakarta and had actually mixed the explosive ingredients.

Such indifference to the suffering of the victims seems to indicate not only rage about the alleged abuses of the West, but a kind of defiant racism. There are subtle differences in the way people are described physically in Indonesian from the culturally neutral '*orang putih*, literally 'white person', to the more controversial '*bulē*', which means 'albino'. The Indonesian version of 'albino' can be used relatively neutrally and often pops up in humour, but when applied dismissively can convey notions of inferiority. It is difficult to tell from a written version in which way the term was used since tone of voice and expression may alter the meaning. In an interview with the journalist, Sarah Ferguson on 23 May 2003, Amrozi is once again recorded making apparently dismissive comments about whites, but in the translation the word 'whities' was used and it is hard to be sure what was actually said in Indonesian. Significantly, the terms for tourist, *turis*, does not appear to have been used in these court cases, which suggests that it was the victims' Western or white attributes that caught the attention of the bombers. Westerners do not of course have to be people of white European descent since not only are most Western nations multi-cultural; there are also Asian nations, such as Japan, that are widely seen as Western.

Amrozi's hatred of Westerners may well be derived from his experiences in Malaysia, but he appears to have been become radicalised through attending a *pesantren*, traditional Islamic college. He admitted to attending the Lukman Nul Hakim College in the 1990s in Malaysia where Abu Bakar Ba'asyir was one of the lecturers, but it is unclear what he actually learned there. Amrozi's hatred of Westerners may have been underpinned by radical Islamic teaching, but some of the other bombers expressed their motives a little differently. Imam Samudra, for example, who has a number of assumed names, expressed his hatred of the victims in religious terms. According to the police he is an engineer with a university education who learned the art of bomb making in Afghanistan. During his trial the prosecutors argued that he chose the targets and led the planning meetings and that he stayed behind in Bali for four days after the bombings allegedly to monitor the start of the police investigation. Imam Samudra is also suspected of being involved in a series of church bombings across Indonesia. When Imam Samudra gave evidence at the separate trial of Abu Bakar Ba'asyir he said that the bombings were part of a jihad, though he denied any connection with the militant group, Jemaah Islamiah. In his response to a question about the Christians who died in those attacks, he said 'Christians are not my brothers'.

Imam Samudra wrote a 280 page book when he was in prison, published in 2004, entitled *Aku Melawan Teroris* (I Oppose Terrorism), in which he cited the Koran to legitimise his attacks and a *jihad* in Bali. He reaffirmed that his target was the USA and its allies, who according to his interpretation of the Holy Scriptures could be killed wherever they could be found. After doing random research, he found that Bali had the biggest homogenous target of Americans and their allies at the Sari Club and Paddy's Pub (2004, 120). In his book, he refers the Americans and their allies as 'nations of Dracula'. Referring the large numbers of Indonesian killed in the attack, Imam Samudra said it was 'human error' (English is in the original). 'And in fact, the human error was what I've been very much regreting' (Dan sesungguhnya *human error* itu sesuatu yang amat sangat kusesali) (2004, 121).

Audiences around the world may have been stunned by Amrozi's reaction to the news of the death sentence, but there was a repeat performance when Mukhlas, the leader of the alleged bombers, responded with apparent delight to the death sentence that was handed out to him by the Indonesian court in the Balinese capital of Denpasar. He became the third terrorist in Indonesia, along with his younger brother, Amrozi and the operation's mastermind, Imam Samudra, to be sentenced to death for inspiring his followers to slaughter Westerners, supposedly to avenge the oppression of Muslims. His antics mimicked the ecstatic reactions of his brother who had been sentenced earlier and who had claimed that there were many people in Indonesia willing to take his place should he die. The bombers may have responded differently to questions about their motives, but one who has offered a clear political explanation is Mukhlas. He was not only the eldest and most experienced of the three brothers, but was a veteran of Afghanistan where he claimed to have met Osama bin Laden.

> Osama bin Laden. Yes, I was in the same cave as him for several months. At the time, he wasn't thinking about attacking America. It was Russia at that time (NineMSM, 23 May 2003).

Sarah Ferguson recorded an interview held in prison nineteen months after the bombing and provided the following translation of Mukhlas's reasoning:

> I want the Australians to understand why I attacked them. It wasn't because of their faults, it was because of their leaders' faults. Don't blame me, blame your leader, who is on Bush's side. Why? Because in Islam, there is a law of revenge (NineMSN 23 May 2003)

This may of course be a post event rationalisation given the bombers' earlier claims that they were attacking Americans, but could equally be taken at face value.What is significant about Mukhlas and his followers is that once charged they did not seek to deny that they had been involved in the killings even to the extent of correcting the judges to make sure that the record was accurate. With the exception of Ali Imrom, who confessed that the bombings had been against his Muslim teachings, Mukhlas's group took pride in their achievements. To draw attention to their religious affiliation, they dressed in Muslim style clothes, wearing sarongs and fez-like hats, during the proceedings and were photographed carrying out their devotions. Unlike the other defendants, Ali Imron wore a suit and tie in court, behaving politely and expressing remorse and even crying a couple of times in public. While in police custody, Ali Imron also gave a press conference about how the bombs were made, displaying a filing cabinet similar to the one that was used in the explosions and other related equipment. He also said that there was no reason for Indonesians to doubt his team's ability to make the weapons and explicitly rejected the view expressed in some quarters of the local media that this was the work of foreign nationals.

The bombers robustly defended their actions and insisted that they were taking part in a *jihad*, a struggle to establish the law of God on earth and a 'Muslim holy war against unbelievers' (*The Little Oxford Dictionary* 1996, 344). The Jihad is sometimes called the 'sixth pillar of Islam' and has two meanings. First, there is what is called the 'greater jihad', a striving of any kind, particularly a moral one, such as the struggle to be a better person, a better Muslim, the struggle against drugs, against immorality, and against infidelity, etc. Second, there is the holy war itself, which is undertaken when the faith is threatened and only with the approval of a religious authority in accordance with Islamic law, Shari'ah (faculty.juniata.edu/tuten/islamic/glossary.html). The Bali bombers do not appear to have had such authority to carry out their attacks and after considering his situation Ali Imron confessed in court (15 September 2003) that the bombers had breached the terms of Jihad, an account that contradicted that of Imam Samudra's reasoning. His misgivings may be summarised as follows:

1. The bombers' targets were not clear, whereas in accordance with the terms of Jihad the target must be clear and there must be authentic proof that those targeted are truly enemies of Islam.
2. There was no warning before the attack, whereas under the terms of Jihad a warning or *dakwah* is required before the attack.
3. The killing of women is banned under the terms of Jihad unless the women concerned have taken up arms against Islam.
4. The bombings involved a very nasty form of killing, whereas in accordance with Jihad any killing has to be done in the best possible (most humane) way.

Ali Imron's conclusion was that the bombings had not been carried out in the true spirit of Islam, ' In addition to breaching the terms of Jihad, acting in accordance with the history of our Islamic predecessors, I found that they conducted Jihad in the manner of the Bali bombings, but in fact with dedication (*beramal*) we must follow them'. He went on to say that '...whatever the motive behind the Bali bombings, the act was wrong because it breached the rules'. Initially, the discussion on the meaning of *jihad* was fairly limited in Indonesia, but this changed after the 1 October 2005 Bali bombings as the VCD of the suicide bombers' confessions were circulated and religious leaders felt compelled to comment. The majority of religious leaders in Indonesia rejected the use of suicide bombings and argued that the instigation of Jihad was only acceptable when the nation was under attack. They argued that Jihad was not acceptable in Indonesia because there was no national threat, unlike the case in Iraq.

Why Bali?

Since the strife of the mid-1960s, Bali had been the safest island in Indonesia, and remained largely unscathed by the upheavals of the Asian Crisis. Bali's comparative security and prosperity may have provoked envious reactions from elsewhere in Indonesia, particularly in areas of great economic hardship, including the big cities of neighbouring Java. Many of the islanders became overconfident about the island's serenity and its destiny, encouraging many to believe that theirs was a 'sacred island', protected by God from evil. This outlook had been reinforced by an event that took place in the early 1980s when a bomb from Java that was destined for Bali exploded on a bus before it reached the island. The fact that there had been riots and bombs in other parts of Indonesia while Bali remained safe may have led the security forces and community in Bali to be overconfident.

Bali probably became a target because the island has become a readily recognisable symbol of global tourism, one of the best-known tourist destinations in Southeast Asia. The reasoning is that whatever occurs in Bali, let alone that which involves Westerners or Western interests, will receive high publicity. Bali's prosperity and security may have made it a tempting target, but what seems to have made it irresistible was that it was a soft target. As a result of the strife that followed the fall of Suharto, security measures in Java and elsewhere had been undertaken to protect embassies and government institutions, turning them into hard targets.

Tourists are not only soft targets but also have the advantage of following predictable behaviour patterns and a tendency to cluster. Tourists are also invaluable because ordinarily there is less backlash to attacking tourists than to indiscriminate bombings which produces more 'innocent local victims'. Bali was moreover not an average destination, but an especially well known one, the epitome of global tourism. The presence of large numbers of Westerners meant that any major disruption would attract Western interest and thus publicise the terrorists' cause. The involvement of foreign nationals would not only attract attention, but would also create publicity that the government could not suppress. Bali's global profile, its openness to foreigners and its lack of tight security made it an especially tempting target. Interestingly, what has emerged from the trials of the Bali bombers is that tourists per se were not supposedly the intended victims, but Westerners and possibly Christians.

Bali is moreover located in Indonesia, which up until the Bali bombings had been regarded as the 'weak link' in the war against terrorism in Southeast Asia. With its diminished state capacity, political and economic weaknesses combined with the unresolved issue of the role of Islam in national politics, Indonesia had become an attractive target for Islamic extremists (Rabasa 2003, 37). In the early years of Indonesia's new and reformist democracy, militant Islamic groupings were able to exert greater influence than the numbers of their followers would appear to justify. A combination of a lack of mobilisation by Muslim moderates and complacency in government enabled radicals to exploit Islam for their own purposes and to operate with relative impunity. The bombings, however, helped to change the political climate prompting more secular politicians and moderate Muslims to challenge the radicals and to support a crackdown on extremists, but whether this newfound resolve will be sustained in the light of ongoing extremist activity remains to be seen.

The 2005 Bombings

On 1st October 2005 Bali suffered a second round of attacks when cafes along Jimbaran Bay and Kuta were targeted by bombers, leaving 23 dead including three suicide bombers, most of whom were Indonesian citizens (see tables 10.2 and 10.3). The first was at Raja's Restaurant in Kuta Square which was bombed at 7.45 pm local time, followed a minutes later by two bombs blasted at cafes along Jimbaran Bay, south of the Bali international airport. They killed fewer people than the attacks of 2002, but the bombs they used were more sophisticated as they contained ball bearings, some of which were found in the bodies of injured victims.

Table 10.2 The Victims' Countries of Origin in the 2005 Bali Bombings

No	Country	Number of dead victims
1	Japan	1
2	Australia	4
3	Indonesia	15 + 3 suicide bombers
Total		23

Source: Sanglah Hospital, Denpasar

Table 10.3 Countries of Origin of the Injured Victims in the 2005 Bali Bombings

No	Country	Number of injured
	Indonesia	102
	Australia	29
	Korea	7
	Japan	4
	America	4
	Germany	3
	Belgium	1
	French	1
Total		151

Source: Indictment against Mohamad Cholily (a suspect of Bali Bombings II) prepared by Bali Prosecutor Officer, 2006, 8.

It took police only two days to announce that the attack was the work of suicide bombers. The police reached this conclusion after viewing a video recording by an Australian tourist, who was with his family and a friend and just happened to be outside Raja's Restaurant photographing the nightlife of Kuta, when one of the bombers struck. He accidentally filmed a man with a backpack who was walking faster than other people entering the restaurant seconds before the attack. General Made Pangku Pastika, the Bali police chief who was the former chief of a multinational investigation on the Bali Bombings 2002, showed the journalists in a press conference held in Kuta how a suicide bomber carrying a backpack could be seen walking through guests having dinner in the restaurant seconds before the bomb went off. Following the relatively rapid identification of the victims, the police were left with the body parts and heads of what turned out to be the three chief suspects. The police thought that by circulating posters with colour photographs of the faces of the suicide bombers they would quickly be able to identify them, but it did not work out like that. Some weeks passed with little progress towards the identification of the bombers being made and so the pictures were revised and clarified by removing the blood and debris on their faces. More posters were circulated, but there was still little response from the public. No family members stepped forward to admit that the bombers were their relatives and the Indonesian public in general and the tourism industry in particular started to become very anxious. The police came to realise that ordinary people simply could not recognise the dead suicide bombers, as opposed to being unwilling to identify them, and they decided to change their approach. Using other sources of intelligence the police, followed by the media, initially alleged that the bombs were the work of two Malaysian fugitives from the Bali bombings of 2002, Azahari and Noordin M. Top, but later had to admit that perhaps it was the work of a new generation of bombers. As the investigation became prolonged, ordinary Indonesians began to worry that the bombers would never be identified and that there might be more attacks planned.

In contrast to the 2002 investigation of the Bali bombings of 2002, which was rather open to the media, the 2005 Bali bombings enquiry was more secretive, perhaps because of fears concerning the bombers' global connections. There were daily and frequent press conferences held by police and journalists, but the public received no significant information on those responsible for the 2005 bombings. The police usually showed up at each briefing to say that there had been no significant developments, though there was a great deal of police activity with over seven hundred witnesses being interviewed. As the investigation continued a special detachment of police launched covert operations, shaking out alleged Jemaah Islamiyah suspects throughout Java. No arrests were announced until after the storming of a safe house in Batu, near Malang in East Java, in which Azahari and one of his followers was killed. Police found dozens of vest-bombs, VCDs, books, and a plan for a 'bomb party' toward the Christmas and New Year, but Noordin remained on the run and by early 2007 had still not been apprehended.

Sidney Jones, the Southeast Asia Project Director of the International Crisis Group has reported that Noordin Top now calls his splinter grouping 'al-Qaeda for the Malay archipelago', although he still regards himself as the leader of JI's military wing. Sidney Jones says Noordin and the people around him are committed to following the al-Qaeda tactic of attacking the US and its allies wherever they may be found. Since JI is based in Indonesia, then neighbouring Australia, one of the USA's prominent allies, remains a target (Radio Australia, *AM*, 6 May 2006). The funding to mount an attack could well be provided by al-Qaeda, but there are also other possible sources, including efforts to raise monies by the terrorist members themselves. For example, prior to the Bali bombings of 2002, some of Imam Samudra's men robbed a gold shop in West Java to help to fund some of the costs of the intended attack.

These costs would include an estimated Rp 3–4 million to make a vest bomb, the rental of premises and the costs of surveying the target.

Papers found at the scenes of the bombings, and in the hiding places of those in custody, revealed details of the October 1, 2005, attacks. The notes show how JI members traveled to Bali to survey potential targets before reporting back to JI's master bomb maker, Azahari. They inspected nightclubs, temples, shopping areas, sports venues, fast food outlets, souvenir shops and the international airport. They homed in on Jimbaran Bay, the eventual scene of two attacks, as a potential target because – 'Insya Allah' (God willing) – they estimated that there would be at least 300 people there when they attacked (Wockner 2006a). Mohamad Cholily (aged 24), one of four suspects of the second Bali Bombings, said that he was with Dr Azahari when they heard news of the carnage on BBC Radio and that Azahari had shouted, 'Allahu Akbar' (God is Greatest) and 'Our project was a success', he said; but in court, Cholily withdrew his statement. Cholily was being tutored in bomb making by Azahari, the so-called 'demolition man' and it was he who led police to the safe house in East Java where Asia's most wanted man was in hiding (Wockner 2006).

Despite the death of the chief bomb maker the public remained fearful, not least because of the discovery of the plan for the 'bomb party'. The vest bombs and other paraphernalia had been confiscated by police, but it was widely held that the Azahari group must already have recruited dozens of people willing to carry out further suicide missions. What also raised tensions was the recovery of video footage containing the pre-recorded confessions of the three suicide bombers in Bali: Salik Firdaus, Aip Hidayatulah, and Misno. Circulated widely in Indonesia and abroad, the confessions appeared to suggest that further attacks were possible. In response to the perceived ongoing terrorist threat, the Australian government issued further travel warnings on Indonesia, including Bali, leading to a renewed decline in visitor arrivals, but there were significant differences as compared to the 2002 attacks. This time there was no massive exodus of tourists under the full glare of the media in the days or weeks following bombings as had happened in 2002. Initially, tourism appeared to be less adversely affected than before, but what made the arrival numbers plummet were the combination of the travel warnings and the televised confessions of the suicide bombers.

Media coverage in Asia and Australia of the terror attacks and the ongoing hunt for terrorists in Indonesia continued to frighten tourists away from Bali. Australia continued to issue travel warnings of the possibility of further terrorist attacks in Indonesia, which was understandable given that Noordin Top was still on the run. Police killed two suspect terrorists and arrested another two during a second raid at dawn on 29 April 2006, in Wonosobo, Central Java, but Noordin Top evaded capture once again, leading to speculation that the number of his followers was in decline and that his capacity to launch further attacks had diminished.

The 2005 bombings combined with perceptions of a global terror threat began to have a marked impact on travel arrivals in Indonesia. During the 2005–2006 Christmas and New Year period, usually the busiest season of the year, hotel occupancy fell below 40 per cent. Widely anticipated as the year in which a full recovery would be made with a resumption of the old the upward trajectory, 2006 turned out to be a disaster with hotel occupancies slumping to 30 per cent. Air Paradise International (API), the Bali-based and -owned airline, mothballed its service totally by 24 November, and was forced to lay off 350 of its employees, some of whom were the airline's Australian employees. Among those badly affected were tour agencies deprived of customers. Garuda Indonesia reduced the frequency of its flights from 32 to 25 services per week between Bali/Indonesia and various cities in Australia and their services between Bali and Japan declined from 22 to 16 times a week. This occurred because of the drop in passenger demand from Bali's two main sources of tourists: Australia and Japan. In the months following the 2005 bombings the arrival figures had more than halved from around 4,500 per day to 2,000 per day, causing Qantas and Australian Airlines to also reduce the frequency of their flights to Bali. By mid 2006, the data were indicating that the impact of the second round of bombings would be as bad, if not worse than that of 2002 (*Kompas*, 11 January 2006, 35). The targets may have been America and its allies, but by targeting tourism great suffering was inflicted on Indonesian people and the Indonesian economy.

As the trials of the 2002 suspected bombers unfolded, a variety of reasons were given for why Bali was attacked, ranging from a simple desire to hit back at Westerners for their supposed attacks on Muslims, to a more politically sophisticated assault on John Howard's support for President Bush, and Australian intervention in East Timor in 1999. Some of the explanations given by the defendants have been couched in terms of what appears to be racial hatred, though these threats and statements are somewhat vague. What has emerged clearly is that they decided to bomb a tourist resort because it offered a relatively easy target with the promise of maximum casualties. The intended victims were not targeted because they were tourists per se, but because they included large numbers of foreigners whose deaths would attract publicity to the terrorists' cause. The bombers expressed some disapproval of the alleged behaviour of tourists in Indonesia, but it was the intended victims' nationalities and perhaps racial type, their invaluable foreignness, that appears to have been why they were targeted. The killing of foreigners is useful because it generates more publicity than when only locals are involved and such publicity is doubly valuable because it is difficult for governments to suppress, enabling terrorists to make their causes known globally. There is also less backlash to attacking tourists than to indiscriminate bombing, which produces more 'innocent victims', and Bali seems to especially attractive because any local victims would be likely to be Hindu and not Muslim. As it happened, the bombers' plans backfired and they ended up killing significant numbers of their co-religionists.

Figure 10.3 T-Shirt outside the Paddy's Bar bombsite in 2003 expresses defiance

Figure 10.4 Indonesian tourists passing the Sari Nightclub bombsite in 2003

Chapter 11

The Rise and Fall of T

For more than thirty years tourism had a strong upward traj ... Bali, shrugging off the occasional minor setback and when a serious downturn came in 2002 and 2006, the islanders were left in a state of shock as their economy was ravaged. This chapter examines the impact of the Bali bombings on international visitor arrivals in Bali and compares this crisis with previous crises to map the rise and fall of tourism since its introduction in the 1920s. Given the industry's importance to the island's immediate and long-term development, there is clearly a need to see how tourism responds to crises over time and to assess recovery rates. The purpose here is chart Bali's development cycle with reference to a very widely used and renowned model proposed by Richard Butler, which is known as the Tourism Area Life Cycle (TALC).

By charting the ebb and flow of tourism over time, we are able to demonstrate that the Bali bombings had by far the greatest impact on international tourism visitation than any other crises in the island's history. Such was the severity of the decline in Bali that special measures were undertaken to restore confidence (see Chapter 9), but, important though these measures were in helping to revive tourism after the 2002 and 2005 bombings, they do not fully account for the strong resurgence in international arrivals that followed. What is suggested here is that the destination had not yet reached its full potential and that the strength of the resurgence owes much to the underlying trend of the development phase from the late 1960s onwards.

On top of the tragic loss of life brought about by the bombings of 2002 and 2005, Bali experienced a dramatic fall in visitor numbers, which was compounded by the war in Iraq and the outbreak of SARS in Southeast Asia. This was not, however, the first time that the island's image has been blighted by strife, but it was clearly the worst downturn in living memory. The purpose here is to compare the impact on visitor arrivals of the Bali bombings and the various other crises that have shaken the island since tourism was first introduced over eighty years ago.

Visitor arrivals are a crude measure of a destination's viability, not least because the means by which the Indonesian authorities compile the statistics are not invariably reliable, but they do provide insights not obtainable by other means. The collection and analysis of these kind of data is also compounded by the fact that record keeping did not begin in earnest until 1969, some forty years after the island was established as a holiday destination. Given the difficulties, this chapter explains how it was possible to plot visitor arrivals over many decades, despite some vagaries in the figures, and to uncover the underlying trends in the island's development. A long-term view is adopted in the manner originally proposed by Butler and it is argued that by comparing the impact of the bombs against the destination life cycle a more revealing analysis is arrived at.

LC

m destinations are dynamic and they change and evolve over time, and this olution is brought about by a variety of factors including changes in fashion, changes in the financial circumstances of the visitors, and changes (or even disappearances) in the natural and cultural attractions that originally attracted tourists (Butler 2006, 4). As facilities improve and awareness develops more visitors will come, and then with marketing, further dissemination and more increases in facilities the popularity of the destination's popularity will surge. Subsequently, the rate of increase will shrink as the capacity of the destination to absorb more visitors is reached. In order to help assess this dynamic evolution Butler (1980) developed a hypothetical model of the development of a resort area by plotting the number of tourists over time. It was based upon the product cycle concept, whereby sales of a particular consumable proceed slowly at first, develop a rapid rate of growth, stabilize and eventually decline. In accordance with this model tourists will visit a destination in small numbers initially restrained by difficulties of access, facilities and local knowledge (Butler 2006, 4).

The product life cycle concept follows an asymptotic curve and when applied to tourism Butler argued that there were six evolutionary stages: exploration, involvement, development, consolidation, stagnation and rejuvenation or decline. In the initial phase no specific facilities for tourists exist but by the involvement stage facilities are largely provided by locals, but local involvement and control declines rapidly in the development phase. At this level locally provided facilities are '...superseded by larger, more elaborate and more up-to-date facilities provided by external organizations, particularly for visitor accommodation' (Butler 1980, 8). It will almost certainly be necessary for regional and national engagement in the planning and provision of facilities and this may not be completely in keeping with local preferences.

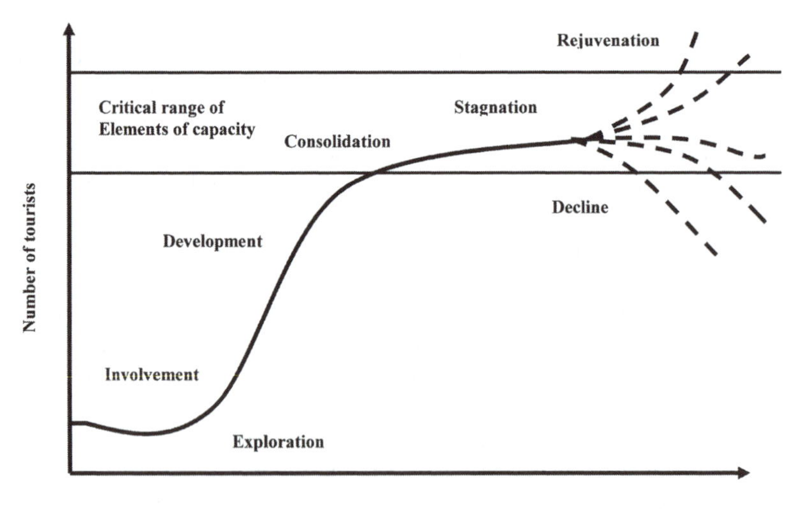

Figure 11.1 Butler's Tourism Area Life Cycle Model (1980)

Butler's model has been revisited over 51 times in the published literature, generating a substantial debate, and many of the papers that have found the model useful refer to research in such diverse settings as the United States, Sri Lanka, Malta, Isle of Man, the Algarve and Australia, but, up until the publication of a paper by the authors of this book in 2006, not Bali (Darma Putra and Hitchcock 2006). The latter publication could well represent the first attempt to understand the dynamics of tourism in response to a highly significant contemporary threat, namely terrorism. A complete account of the debate on the TALC lies beyond the confines of this chapter, though the authors are aware of Berry's table (2001) and Lagiewski's literature survey (2006). The important point is that, often with some modification, the model has proved useful in diverse contexts and as far as this analysis of Bali is concerned, perhaps the two most relevant attempts to develop the model have been by Strapp (1988) and Johnston (2001). Using a case study from Sauble beach, Ontario, Strapp proposes using the average length of stay to calculate the total person days, as opposed to visitor numbers to accommodate the change of status of tourists who eventually retire in destinations. In Strapp's work the emphasis was on the increasing use of tourism resorts as retirement centre, which has in the last decade become an important feature of Bali's tourism industry. For example, the numbers of retired people from the Netherlands taking holidays in Bali has risen in recent years but the precise figures on retirees in Bali are difficult to obtain. The collection of data on length of stay in Bali is still also irregular and thus Strapp's modifications cannot be applied at this stage, though the potential for future research is evident.

In contrast to Strapp, Johnston (2001) takes a more general view and without linking his analysis to any particular region in the first place he has tried to strengthen the model's theoretical foundations. Using a modified form of 'grounded theory' Johnston came up with a re-drafted version of the life cycle with two notable differences from the original. First, the original model was divided into three eras, which Johnston concluded were based on a basic geographical process, to which he added the following comparative variables: human life cycle, product life cycle, port development and eco-succession. Johnston thus proposed revising the life cycle to include: a pre-tourism era, from before the exploration phase and extending through into the involvement phase: a tourism era, which could lead on to the mature epoch and even include a post-stagnation phase; and a post tourism period when a new institution prevails at the local level. Johnston replaced the growth in visitor numbers with growth in accommodation units, and applied this model to the case of Kona in Hawaii (Johnston 2002). Given the expansion of the accommodation sector in Bali, and the significance of local ownership, Johnston's observations are relevant, and, although the overall statistics are not reliable, they are improving. According to the statistics provided by the Bali Tourism Authority (2003) the total (star, non star, and hostel) was 29,754 rooms in 1998, but by 2002 it had increased to 35,212 rooms. There may come a stage when Johnston's scheme could be applied in Bali, not least because the housing market, especially for expatriates, is of increasing significance in Bali, whereas hotel occupancy remains much diminished. It may well be that the terrorist attacks have heralded in post-tourism-like developments more quickly than could have been anticipated.

Butler's work may have been held up to considerable critical analysis with some analysts criticising his terminology, not least the use of the term 'stagnation', which has a negative ring, but could have more positive associations including the desirability of preventing further development to conserve cultural and natural assets. In a paper published in 2000, Butler robustly defended his model, arguing that it was still valid, applicable and relevant in every way in the 21st century. An important point to note is that the situation is not static, but is dynamic and that the curve may be revised upwards with sound planning and management. The TALC, however, also remains relevant in the absence of management and control where the situation becomes unstable and stagnation sets in (Butler 2004). Butler also pointed out that the concept of the life cycle in tourism in his 1980 paper represents an early call for sustainable tourism development (Butler 2000, 284–99) and in a earlier paper he argued that, while there were often pre-impact assessments, post-impact assessments were much rarer and needed to be incorporated if tourism were to be sustainable (Butler, 1993, 152).

Here, Butler's concept of the life cycle is applied to Indonesia's most popular tourism destination, the island of Bali, in order to assess the impact of the 2002 bombings on tourism arrivals and the possible speed of recovery. When Butler devised his life cycle concept, the idea that tourism was vulnerable to threats, either environmental or man-made was in its infancy and was not factored in. But by the 1990s (for example Edgel 1990; Aziz 1995; Pizam and Mansfield 1996; Hall and O'Sullivan 1996) papers started to appear on risks faced by tourism, particularly with regard to terrorism, and though the literature is sparse a recurrent theme is that a successful tourism industry requires political stability. A variety of crises had an impact on tourism, but perhaps one of the most important – and possibly least understood – is the threat posed by terrorism. As has been documented in a number of cases (for example Rome and Vienna in 1985) terrorism is a real obstacle to tourism expansion (Salah Wahab 1996, 175–202); terrorism not only reduces tourism activity, but also relocates tourism and has an impact on long-term investment in tourism (Wall 1996, 145). Tourism can, however, bounce back once the threats are removed, as happened in Egypt in the aftermath of the Gulf War (Aziz 1995, 92).

The Collection of the Data

In this chapter the life cycle concept is used to compare the impact of the Bali bombings on tourism arrivals with other crises that have afflicted the island. The statistics on tourism arrivals are based on completed landing cards at Ngurah Rai Airport and have only been collected systematically since 1969 and thus a variety of other published sources are used to garner this information, especially academic and governmental, to provide a picture of what happened before this date. Not to do this would present Bali as a destination where tourism development began in the second half of the 20th century, when in fact the industry is much older and dates back to the 1920s. Taking into account what happened in the early phase of tourism enables us to comment on how tourism development often differs in colonial societies and to contrast it with the general scheme proposed by Butler (1980). The graph has been plotted using statistics covering the period from 1921 to 2004 and an excel file has been used to plot a curve based on the product cycle concept in which the number of visitor arrivals are plotted against time and an account of the sources used and a critical appraisal of their reliability are provided below.

Statistics on visitor arrivals began to be collected systematically in Bali in 1969 and have continued up to the present. Completed landing cards are collected from international visitors arriving at Ngurah Rai Airport and the Bali Tourism Authority (BTA) regularly publishes the data. There is, however, a time lapse between the handing in of the cards and the publication of the data and thus it was not possible to have a completely accurate figure of the annual impact of arrivals of the 2005 bombings at the time of publication, though some figures are provided and show once again a considerable drop. In the BTA's publications the data are listed as an annual overall total and on a monthly basis; the tables also show the main countries of origin of tourists: Australia, Japan, France, Germany, the Netherlands, United Kingdom, USA, Taiwan and Korea. What the statistics do not reveal are tourists arriving from elsewhere inside Indonesia and thus domestic tourists are overlooked, as are foreign tourists who originally entered the country at another airport. The statistics also omit the tourists who arrive by sea from other Indonesian islands and from cruise ships. The figures moreover do not differentiate very clearly between visitors arriving for recreation and visitors travelling as tourists whose real reason to visit Bali is business, and whose numbers are substantial. In the absence of additional ways of accurately recording visitor arrivals, the number of visitors passing through the international terminal at Ngurah Rai airport remains the best available source of statistical data.

In order to assess the total number of visitors from before 1969, then a much less accurate method has to be adopted. To account for this, an average based on the reasonably informed guesswork of a number of different authors is adopted. Working from a list that was originally compiled by MacRae (1992, xii), the most likely scenario has been adopted, which becomes progressively more accurate in the 1960s. In any case, the number of visitors was so small pre-1969, in comparison to the late 20th century and early 21st century, that changes in arrivals do not alter the overall picture much. The majority of authors concur that the Dutch prepared for the introduction of tourism to Bali shortly after their annexation (1906–1908) of the last independent kingdoms, but that it did not really pick up until the end of the First World War (Boon 1977; Picard 1996). The start date of meaningful visitor arrivals may be taken to be 1921, though the opening of tourism facilities and the marketing of the island began shortly after the pacification of the island in 1914 (Picard 1996, 23).

Picard, for example estimates that annual arrivals in the 1920s comprised only several hundred (1990, 40) and that they rose to several thousand (ibid.) in the 1930s or a hundred a month according to Hanna (1976, 104). The first data published by the Tourist Bureau recorded 213 visitors in 1924, a number that increased steadily to 1,428 in 1929, and after stagnation in the Depression that followed the 1929 Wall Street crash visitation rose to reach 3,000 by the end of the 1930s (Picard 1996, 25). Tourism rose in the 1940s with around two hundred and fifty visitors arriving per month (ibid), but collapsed during the Japanese occupation (1942–45) of the Dutch East Indies in the Second World War. Arrival figures from the occupation period are not accessible, but as Yamashita notes (forthcoming), the Japanese did organise tours to the territories that they conquered; tantalizing glimpses of how Bali's tourism facilities were used by the invaders can be seen in K'tut Tantri's (a.k.a. Muriel Pearson, Manx and Surabaya Sue) controversial book *Revolt in Paradise* (1960) and Lindsey's biography of K'tut Tantri (1997). Lindsey has suggested that K'tut Tantri, the expatriate owner of a beach bungalow hotel on Kuta Beach in the inter war years, may have had some freedom of movement during the occupation and may well have cooperated with the Japanese in the operation of a hotel, possibly the KPM Bali Hotel (1997, 115). K'tut Tantri, however, adamantly denied that she ran a hotel for the Japanese, and claimed that her own hotel was demolished by Balinese eager for materials. What really happened remains unclear, but it seems likely that the Japanese forces continued to use Bali's tourism infrastructure, notably hotels, presumably for recreation purposes and as brothels for the euphemistically named 'comfort women'. Recuperating Japanese soldiers may broadly be defined as tourists, though they may not conventionally be thought of as such, and thus the occupation period is included in the Bali life cycle diagram, though the precise numbers remain speculative.

There was a small tourism industry throughout the late 1940s and 1950s, but data on visitor arrivals are equally hard to obtain, but by the 1960s the Indonesian government had started to compile the statistics, initially for Indonesia as a whole, but for Bali in particular by 1966 (Udayana and Francillon 1975, 723). It would be tempting to ignore the pre-1969 data because it is not especially reliable, but this would miss out something significant about Bali, its long period of early development that equates with the exploration phase on Butler's model.

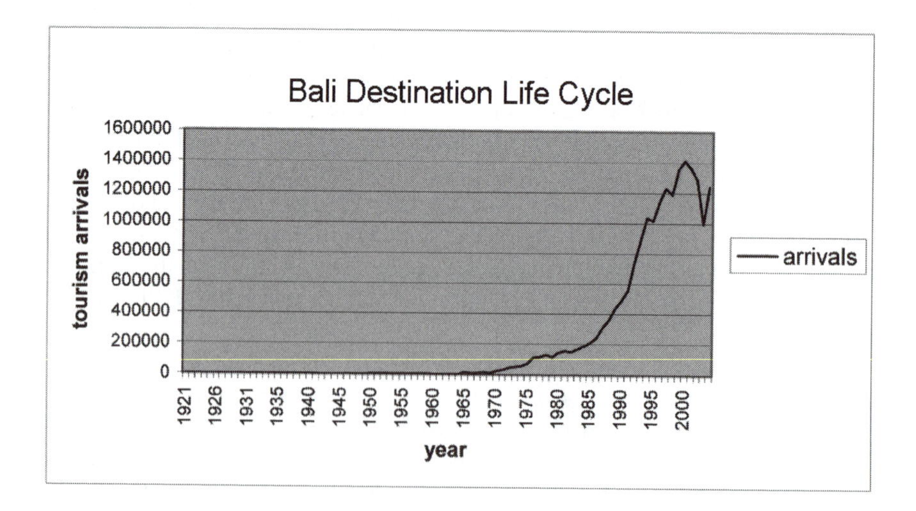

Figure 11.2 Bali Destination Life Cycle

The Bali 'S' Curve

The Bali TALC or 'S' curve was first plotted in 2005 before the second round of bombings and was published in *Progress in Development Studies* the following year (Darma Putra and Hitchcock 2006). The main purpose was to show what can be learned from comparing the above diagram with Butler's original model and to demonstrate how underlying trends have an impact on the decline and recovery that followed the Bali bombings. A noteworthy feature of the Bali curve is lengthy period of early development that corresponds with Butler's exploration phase. The longevity of this phase, lasting more than forty years, may be partly due to the variety of governments experienced in this period: Dutch colonial (1908–1942), Japanese Imperial (1942–45), the attempted restoration of colonial rule (1945–49), the proclamation of Indonesia's independence and the establishment of the republic (1945 onwards). During the exploration period it was the colonial regime that inaugurated the development of tourism facilities through the KPM (Royal Packet Navigation Company) by operating passenger ships and developing the first hotel on the island, the KPM Bali Hotel, which opened in 1928. In striking contrast to Butler's model, it was outsiders who took the initiative and locals did not become seriously involved until the 1930s. Even when locals did become involved it was often in association with a foreigner, as was the case in Ubud where the Russian-German artist, Walter Spies, built his accommodation facilities with the help of the Cokorda Raka Sukawati, the local ruler. A significant local involvement was the participation of Dutch-sponsored Balinese dancers and artists in the Colonial Exhibition in Paris in 1931 (MacRae 1992, xi). Parallels can be drawn with other colonial societies, notably the Pacific where external investors played a leading role in tourism from the outset (Harrison 2003, 3–12) and it is suggested that these kinds of contexts represent something of a special case that do not detract generally from Butler's observations.

The successor regimes of Sukarno (1950–1966) and Suharto (1966–1998) that came after the Dutch withdrawal, continued to support tourism development, and locals began to be more actively involved in opening small scale handicraft businesses and providing accommodation, first in Sanur and Denpasar in the 1960s, and then Kuta in the 1970s. Central government support was crucial since without it tourism would not have moved much beyond the early exploration phase. President Sukarno, for example, opened the Bali Beach Hotel in 1966 and started the expansion of Ngurah Rai Airport, which eventually opened in 1969 under Suharto's rule.

The case of Bali does not precisely resemble Butler's model for a variety of reasons that need to be contextualised with regard to local social and political conditions such as the building of the Bali Beach Hotel, along with three other hotels elsewhere in Indonesia (Hotel Indonesia in Jakarta, Pelabuhan Ratu in West Java, and Hotel Ambarukmo in Yogayakarta, Central Java) with funds provided by war reparations from Japan. As tourism began to be developed more rapidly, local ownership began to be a strong feature of Bali's tourism, not only in accommodation, but also in entertainment and souvenir production (Vickers 1997). Regional and national government played an increasingly strategic role as tourism moved into the development phase in Butler's model, but in Bali the government was active from the outset. Again in contrast to Butler's scenario, local ownership expanded in medium and small-scale ventures as tourism developed and did not give way to foreign investors. Foreign investors did, however, eventually become heavily involved and the bulk of this investment went into the resort of Nusa Dua in the 1970s. As with the Bali Beach Hotel earlier, it was not simply a case of big investors becoming interested since Nusa Dua was also the recipient of funding from international development agencies such as the World Bank and the Asian Development Bank. National and international investors also became involved in related sectors, and not just hotels and aviation, including shopping malls, cruise ships and export agencies.

The Bali TALC is based on statistics that fluctuate over time and is thus not smooth like Butler's schematic version, but the wobbles are interesting when tied to local and global events. The Gulf War had an impact of visitor arrivals early in 1991, but the recovery in the remainder of the year rapidly replaced earlier losses and has little impact on the graph. This was also because Bali had the geographical advantage of being far from the conflict in the Middle East and relatively close to its important markets in Australia and Japan. In addition, the selection of Bali as the venue for the Pacific Area Tourism Association Mart and Conference, held in Nusa Dua in April 1991, helped to raise the profile of the island as a peaceful destination. After this event, Bali received more international airlines both in term of number and frequency, predominantly from Asia, Australia, and Europe.

The Gulf War had little impact on annual visitor arrivals, but something closer to home did: the outbreaks of cholera amongst returning Japanese tourists in 1994 and 1996. The Japanese comprise Bali's second largest source of tourists and the small dips on the graph can reasonably be attributed to the fall in Japanese arrivals in that period. There is another notable dip, however, aside from the 2002 Bali bombings, which occurs in the late 1990s and is attributable to be the Asian Crisis (1997–98), which accelerated the fall of Suharto and led to a vote for independence in East Timor under President Habibie. The turmoil did not badly afflict Bali, but restrained the strong growth in tourism arrivals in the 1990s (Hitchcock 2000). Reassured by the relative security of Bali as compared with the rest of Indonesia and low prices occasioned by the fall in Indonesia's currency, the Rupiah, Australians returned in larger numbers than before. But the upturn did not endure because Indonesia experienced both ongoing political instability and the well-publicized strife that accompanied the independence vote in East Timor. By 2000, however, with the Timor Crisis over and a newly elected president in office visitor numbers surged once more.

The most noteworthy feature of the graph is the huge drop – vastly larger than any previous crisis – that followed the 2002 Bali bombings. This rapid decline resembles the Egyptian experience of 1991 when tourists were perceived as being directly threatened and stayed away, though as with Bali numbers rose again in less than a year. Of the two, Bali's recovery is perhaps the more remarkable given that the Bali bombs were followed by fears about the Second Gulf War, SARS, and Bird Flu, not to mention more bombings in Indonesia. In particular, there was the bombing of Jakarta's Marriott Hotel, a combined tourism and business travel facility, and this had the potential to undermine Bali's recovery. The recovery from the bombings crisis was also threatened by the introduction of visas on arrival in 2003 in the country and this caused outrage among Indonesians with involvement in tourism since the country had enjoyed a twenty-year-old open door policy. In order to boost arrivals, President Suharto had scrapped visas for tourism and business travel in 1983, less than one year after the opening of the first hotel in Nusa Dua.

It was, however, not all gloom and doom since the PATA Conference passed off successfully in Nusa Dua in April 2003 and some of the shortfall in overseas visitors was met by a rise in domestic tourism. Indonesia's then president, Megawati Sukarnoputri, tried to alleviate the crisis by relocating ministerial meetings to Bali, and through the provision of a budget to support a range of other conferences and receptions. Perhaps the biggest boost was provided by the arrival of all ASEAN leaders for the ASEAN Summit, the participation of Australian Prime Minister John Howard in the first bombing commemoration service and the visit of American President, George W. Bush, all of which occurred in October 2003. The president also amended mid-week holidays associated with religious festivals so that Indonesian citizens could take longer breaks that included the weekend, but once the situation had improved these temporary measures were revoked.

Trans National Corporations (TNCs) are popularly believed to be not especially beneficial for local businesses, but their presence seems to have helped speed up the recovery process due to their access to developed marketing facilities. The presence of such corporations helped to raise Bali's profile internationally in general terms and thus has a knock-on effect in enhancing the sustainability of the island's tourism industry. As is discussed in Chapter 9, the combination of international, governmental and local measures to restore confidence in Bali as a tourism destination provided a united platform for recovery and thus Bali provides an interesting case study in crisis management.

Essential though these confidence-building measures were, they do not completely account for the rapid resurgence of Bali's overseas visitor arrivals less than ten months after the bombings and this is where it is helpful to revisit Butler's model with regard to understanding the crisis in Bali. Before the bombings of 2002 the underlying trend in tourism arrivals was strongly upwards, indicating that Bali was still in the developmental phase of its life cycle had not reached the start of the consolidation phase on Butler's scheme. It is suggested therefore the one of the reasons for the return of a strong growth profile is that the upward dynamic had resumed and was only temporarily derailed by the bombings. If Bali had been further along the curve it is possible that when the bombers struck that it could have necessitated a longer recovery period, leading at best to stabilization or at worst to a longer-term decline.

As the graph plotting visitor arrivals reveals, the 2002 and also 2005 bombings had by far the largest impact on international tourism visitation since the commencement of Bali's tourism industry in the early 20th century, excluding the period of Japanese occupation. There was of course a very severe strife at the end of the Sukarno period in 1965, but visitor numbers were still insufficiently numerous for it to have much impact. Moreover, comments in the visitors' book at the palace in Ubud from this period are resoundingly positive, seemingly unaware that Indonesia was engulfed in a major crisis. Since 1969, when statistics began to be collected systematically, up until the early 21st century there have been a number of major crises, especially the one that followed the Asian Crisis, but none of these are remotely comparable with the Bali bombings.

Admittedly, special efforts were made to restore confidence in Bali as a tourist destination, such was the severity of the downturn that followed the bombings. Marketing by TNCs not only benefited the businesses owned by the TNCs, but also had a knock on effect in terms of a generally improved image that had a positive impact on locally owned businesses. Other confidence boosting measures included high-level conferences, visits by heads of state and even changes in national holidays to stimulate domestic tourism. There was also widespread local support for restraint in dealing with the crisis so that the island did not descend into inter-religious and inter-ethnic strife as economic conditions deteriorated. Vital though these measures were, they do not fully explain the sharp upswing in international arrivals once the security situation had improved. The resumption of a strong profile in visitor arrivals suggests that the destination has not yet reached the consolidation phase in accordance with Butler's hypothesis and that the resurgence owes much to the underlying trend of the development phase associated with the general TALC picture. In a context where tourism development modelling remains inhibited by the lack of comprehensive data gathering, it is not, unfortunately, possible to place a figure on what the underlying trend contributed to the recovery.

Figure 11.3 Sign in 2003 announcing the reopening of Paddy's Bar

Chapter 12

Coping with Globalization

For nearly half a century Bali's development processes and strategies have been primarily concerned with tourism, and this is reflected in the literature and the debates that have emerged. But since the start of the 21st century, in combination with the intensification of democratization and the experience of terrorism, these three issues – tourism, democratization, and terrorism – have not only integrated Bali more closely into national debates, but have made the islanders more aware of the global context in which their economy operates and how their island is perceived. The Balinese are thus compelled to think not only about the benefits of development, but also the accompanying problems, some of which, like the bombings, are very hard to predict

In particular, the bombings of 2002 and 2005 have provided strong reasons for Bali to change its development paradigm or strategy. The Balinese response to modernization and globalization, which are often seen on the island as being two sides of the same coin, has been culturally dynamic, especially in response to tourism and the mass media. The Balinese have taken considerable pride in the use of their culture in defining their regional and ethnic identity, and take it as a compliment that tourists should want to travel so far to enjoy it (Picard 1997; Darma Putra 2003). It is widely held in Bali that tourism and Balinese culture have developed symbiotically through the use of culture to attract tourists, which in turn stimulates cultural creativity. But, the unprecedented nature of the bombings and the fact that they were linked to an international terrorist network profoundly shocked the Balinese. One of the consequences of the attacks was the weakening of the local tourism-based economy, leading to a great deal of soul searching since the island had become heavily dependent on it. To complicate matters the attacks occurred at a time when Indonesia was starting to experience democratization. For many people the excitement of having direct elections in newly radically reformed Indonesian political system has been tempered with fears about social stability. It is widely recognized that peace is a precondition for tourism and that some sort of strategy will have to be enacted to make sure that it becomes a reality.

To complicate matters even further, Bali has become one of the barometers of Indonesian stability in international eyes since any expression of social conflict or unrest on the island will be immediately subjected to international media attention. The bomb attacks of 2002 and 2005 demonstrated just how fast the news was spread and how quickly there was an exodus of tourists and a plummet in visitor arrivals. The case of the bombings may seem an obvious example, but even a minor political-based conflict that involved the killing of political party members in North Bali in the run up to the general election in 2003, provoked widespread national and international media interest, adding to the image that Bali's stability had been weakened. Ironically, it is the ongoing presence of many foreigners living more or less permanently on the island that helps to ensure that news stories receive media attention. Bali's tourism market may have declined, but its real estate market is thriving as expatriates, undeterred by the bombings, seek holiday homes and even permanent abodes.

This chapter examines that solutions offered locally as the Balinese have come to terms with globalization and its implication for development. In particular it focuses on the role of local culture and values in the processes of progress and development since it is often local values that are overlooked or ignored in the interest of progress. Hobart has argued that '...the knowledges of the peoples being developed are being ignored or treated as mere obstacles to rational progress' (1993, 2), but this needs to be qualified with reference to Bali.

Cultural Solutions

In general the Balinese welcomed the introduction of tourism and its attendant developmental benefits, but there has been a long history of opinion formers raising concerns about its potential negative impacts, forcing government officials and members of the intelligentsia to come together to develop what might be called a cultural solution. By the early 1970s, Bali's tourism potential had started to become a reality as arrival numbers soared following the opening of the Bali Beach Hotel in 1966 and Bali's international airport in 1969. Central government started to promote Bali as one of Indonesia's main tourism destinations and in its first five year plan of development (*Repelita*), Suharto's government regime prioritized tourism by inviting international resort consultants and international financial institutions to set up and fund a plan to create a new resort at Nusa Dua. This project proved to be a great success in the eyes of the government as dozens of international hotel groups became involved, thereby enhancing Bali's image as an international class destination. The presence of these major hotel groups also had a knock-on effect on Indonesia's profile in general, as a stable and developing nation.

The opening of the international airport provided easy access to the island for the middle- and upper-income tourists from all over the world who stayed in up-market resorts such as the Bali Beach Hotel, and the other almost luxurious hotels in Sanur. At the same time, budget or backpacker type of tourists also flooded into Bali, usually staying around Denpasar and Kuta and by the 1970s Kuta's development in particular was characterized by local villagers turning their homes into small-scale hotels. As news of the quality of Bali's beaches and waves spread, the former villages of Kuta and Legian rapidly became a 'surfing Mecca' for international surfers, many of whom were Australian. The backpackers and budget travellers were initially welcomed on account of their contribution to locally-owned businesses, but concerns started to be expressed about their cultural impacts. They were judged by their appearance, particularly by their long hair and casual and revealing clothing, and by the habit of some of them of riding motorbikes in town without their shirts. Because of the way they looked they were often referred to derogatively as 'hippies'.

The hippy phenomenon sparked a wide controversy in Indonesia and in March 1971 the Indonesian government announced that visas would not be granted to these dropouts (McKean 1971, 21). In addition, posters showing what hippies looked like and saying that they were barred from offices or other government buildings were put up, as was the case in other parts of Southeast Asia (Vickers 1998, 2). It would appear that the government was concerned that young Indonesians who socialized with hippies, befriending them, acting as their guides or even becoming their lovers, risked becoming corrupted by them and were thus a danger to national morals (Picard 1996, 79, 226). Various reports in the local newspaper the *Bali Post* between the late 1960s and the 1980s portrayed negative images of hippies, referring to them, for instance, as 'nauseating gangs', 'penniless guests', 'drug addicts' and as 'practitioners of free sex' (Darma Putra 2003, 206–7).

Many Balinese, especially those directly involved in tourism, recognized that the so-called 'hippy' was not wholly a bad kind of tourist, but they were very much interpreted as representing a version of Western culture that could bring about negative impacts on the local youth and culture. The government threat not to grant visas never came to reality, presumably because the hippies were too economically valuable to turn away, but such actions reflected the government's and public's growing concerns about the perceived negative impacts of tourism. At the same time, there was intensification of the debates about the so-called over commercialization of Balinese arts and culture. As demand increased for Balinese dances, for which visitors were prepared to pay, the Balinese began to worry that the more sacred performances would end up as entertainments for tourists. In the late 1960s, amidst growing concern, government officers and cultural experts met to draw up guidelines on what type of performance could and could not be performed for tourists. As a result of their deliberations three categories of performing arts were recognized, including the sacred (*wali*) for God, ceremonial dances (*bebali*) for accompanying rituals, and the profane (*balih-balihan*) for public entertainment (Picard 1996, 156). The last type of performance was often simplified and shortened since ordinary tourists, on their own or in groups, did not normally have the time or patience to watch performances that could last for three to five hours.

The so-called 'cultural solution' was given legal status when the island's government issued a regional regulation in 1974 known as the 'Cultural Tourism Regulation' which legally endorsed the concept of cultural tourism. This regulation was revised in 1991 with one of its main aims being the use of Balinese culture to attract tourists and the benefit from tourism to enrich Balinese culture. This policy received international support from agencies such as the French tourism consultants, SCETO, who worked in Indonesia in the early 1970s and also by the former secretary of the United Nations, Javier Perez de Cuellar, who said at the International Conference on Culture and Tourism in 1995 in Yogyakarta, that '… there is no tourism without culture' (*Kompas*, August 24, 1995, 5). With such eminent support the Balinese people and government felt confident that they had chosen the right tourism strategy, and a manifestation of this was the growing popularity of the phrase 'tourism for Bali, not Bali for tourism' among Indonesian officials. The Governor of Bali Ida Bagus Mantra (1978–1988) and the principal initiator of the annual Bali Arts Festival often used this phrase, as did his successor, Governor Ida

Bagus Oka (1988–1998), who popularized it in his period of office in a different way to refer to the tendency to overdevelop Bali when a seemingly unstoppable flood of investors had become attracted to the island. The phrase that he often used to encourage investment was '*membangun Bali, bukan membangun di Bali*' ('develop Bali, not develop in Bali'), by which he meant that the island and the Balinese should be the focus and subject of development not merely objects to exploit.

Another important cultural innovation occurred in the late 1970s in the form of the Bali Arts Festival that aimed at encouraging the development, preservation, and appreciation of Balinese arts and culture. As aptly put by Kagami, the Bali Arts Festival is 'a response to the beginning of mass tourism in Bali in 1970s' (2003, 70). This programme was endorsed by Central Government and was inaugurated by Mrs Tien Suharto in June 1979 on the *puputan* soccer field, the site of the death struggle of 1906. The aim of the festival is not to attract tourists, though many actually do attend, but rather to encourage Balinese people to appreciate their own culture. The month long festival is held annually in the school holiday period, mid-June to mid-July, in order to allow the younger generation to watch and participate and to enrich their appreciation to their own culture. The festival is divided between five distinct areas: the arts parade, arts performance, arts exhibition, cultural seminar, and arts competitions. Old master dancers are invited every year to perform and are given awards for their talent and dedication to developing Balinese arts and culture.

Not everybody was enthusiastic about the decision to hold a festival every year and there was some debate in the local newspapers about the timing of the festival with one group arguing that an annual event would be difficult to sustain. They argued that it would be better to hold it very two to five years to give artists the time to create high quality works, but the majority sided with the yearly festival saying that one year had always been sufficient for new and unexpected developments. Others criticized the amount of money being spent on the festival (currently up to Rp 2.5 billion), but it was more widely felt that the investment in art was worthwhile in cultural terms.

Since it first launched, the annual Bali Arts Festival has been held continuously notwithstanding the social, financial and other crises that have afflicted the island since the late 1990s. President Suharto and his ministers as well as other presidents including President Megawati Sukarnoputri and Susilo Bambang Yudhoyono have all opened the Bali Arts Festival. The Bali Arts Festival has moreover become a model for other Indonesian provinces, and some, like Yogyakarta and West Sumatra, have follow suit but have thus far not been able to host their festivals annually like Bali. Since its inception the Bali Arts Festival has also provided a venue for visiting dance troupes from countries such as Japan, America, Korea, Australia, and India. Their presence adds a global dimension to an essentially local celebration. The Bali Arts Festival has not only encouraged the lively and dynamic development of Balinese arts and culture, but has instilled a sense of pride in local culture.

Not only have foreign troupes visited the Bali Arts Festival, but the Balinese themselves have also been invited to participate in festivals abroad, a trend that can be traced back to 1931 and the Colonial Exhibition in Paris. Bamboo percussion groups from West Bali have, for example been invited to Japan; Balinese artistic residencies have been established in many countries to teach foreign students; in the

USA a Balinese gamelan orchestra has been established combining American and Balinese performers. The Balinese have also a strong profile in the material arts and the price of paintings by Balinese artists has risen sharply, and there is a flourishing export trade in handicrafts and garments. These cultural policies were developed in tandem with the growth of tourism and thus the attacks of 2002 and 2005 were a blow not only to tourism per se, but to the island's cultural fabric. Despite the apparent success of the island's cultural activities many issues remain unresolved such as the ongoing poor remuneration of dancers performing in hotels and restaurant.

In 1998 the provincial government of Bali issued its long awaited decree concerning three aspects of the use of performance in tourism: the assurance of quality of performing arts, list of types of performance that can and cannot be performed for tourists, and the minimum wage for performers. A prolonged debate ensued as the different sectors tried to interpret the rules in accordance from their own perspectives. For example, hoteliers and others working in tourism and hospitality were criticized for paying low wages to Balinese performers, but they responded by saying that they paid generous amounts to the brokers who arranged these performances. Hotels and restaurants avoid making direct deals with groups of performers, because dealing with brokers gives them the assurance that the artists will come at the time required, often at short notice. According to the government's ruling one dancer should receive a minimum wage of Rp 20,000 (equivalent to US$ 2.50) for a two-hour performance. In reality, however, dancers are most likely to spend several hours extra on the job if one includes the journey to the hotel, preparation and make up, the performance itself, and journey back home, and thus US$ 2.50 becomes a meagre reward. Both artists and their employers within tourism initially welcomed the decree, but with no means of enforcing it significant changes remain unlikely. Who for example should be responsible for checking up on whether artists working in so many different locations met the required standards? In addition who would be responsible for monitoring the use and abuse of Balinese religious customs and traditions within tourism? There has been criticism of the custom of marrying foreigners who are not Hindus in accordance with Hindu custom, as was famously the case with Mick Jagger and Jerry Hall, but the practice continues. Concerns were also raised about the use of a Balinese sacred site as the backdrop for a rock video in 2000, but again nothing came of it. Despite such problems many members of Bali's intelligentsia remain positive about the synergy and dynamism of the link between local culture and global tourism.

In 2000 there was a lively debate about the withdrawal of the cultural tourism decree on the grounds that it marginalized the role of customary villages since such local organizations played no rule in determining the development of tourism in their areas (Pitana 2002, 104). The idea of withdrawing the decree was rejected, but the role of customary villages in determining tourism development in their immediate areas was approved. In the aftermath of Indonesia's reformation and devolution of regional autonomy, many villages have begun taking part in managing nearby tourist attractions such as in Tanah Lot, Alas Kedaton, Sangeh, Tampaksiring, Goa Lawah, and Uluwatu. They also now receive a share of the income from ticket sales for the attraction in their area. The argument in favour of retaining the cultural tourism decree was that Bali had no option but to continue relying on cultural tourism (ibid.)

Figure 12.1 **Legong Keraton dance, performed during the Bali Arts Festival, one of the major showpieces of Balinese art and culture**

and it was clear from the outset that coming to terms with cultural tourism would be a balancing act rather than a regulatory issue (Williams and Darma Putra, 1997), though a clearer exposition of the island's cultural policies would help to clarify the discussion.

Economic Solution

The development of tourism in Bali since the 1970s has gradually made the island's economy highly dependent on it. The agricultural sector, which used to make the dominant contribution to the island's well-being, has been slowly marginalized, due among other things to the decrease in the acreage of rice paddies and disorder in the irrigation system, resulting from tourism-related construction and increased housing development. The contribution of agriculture to the regional income in 1971 was 59.1%, but had slumped to a mere 19.81% by 2000. By the same token the contribution made by hospitality over the same period had risen from 33.4% to 62.35% (Erawan 2003, 265). In everyday life, the number of people who directly and indirectly benefit from it has risen, at least until the 2002 bombings, and it is clear that Bali is not only highly dependent on tourism and that this industry is the most significant locomotive in the island's economy.

Tourism undoubtedly makes the major contribution to the island's well being, but its benefits have not been evenly spread and there are gaps between the different sectors, regions and communities (Erawan 2003, 267). In particular this inequality in income has led to regional imbalances with migrants from the poor regions to the east and west being drawn into the already densely populated south and centre. A common pattern is for these migrants to move to town to study and then work, eventually remaining in the urban south to take advantage of the better employment opportunities and educational facilities, leaving many rural villages denuded of people of working age, which also has a knock on effect on agriculture. The number of people who are able to work in wet rice agriculture is in sharp decline, as the children and the younger generation leave their ancestral lands for Denpasar. I Wayan Arthawa has captured the despair that often accompanies this separation in a poem (1994):

Ancestral Land

> Scratching characters on palm leaves
> which poem must flow forth
> to let the mind settle in meditation?
> all the children are fleeing
> leaving their ancestral lands
> destroying the land you live in leaves an empty feeling.
> ....

Without the appropriate agricultural skills the young Balinese would have found it hard to survive in their villagers, but life in the city is no less tough where there is competition between the local (*penduduk asli*) and the migrant (*pendatang*). Less economically able locals who thought that they would not suffer hardship on their home ground have had a rude awakening since the new arrivals have often proved to be more competitive, leading to an income gap between some locals and the more determined migrants. There has been a strong tendency for locals to decide to sell their land for consumption purposes with the migrant buying it and working hard to make the new acquisition become profitable. At community level this tendency is often described in the following phrase:

> ...the migrant sells beef balls to buy land, while the Balinese sells land to buy beef balls.

The phrase simultaneously warns against laziness, encourages the locals to emulate the enterprising behavior of incomers and draws attention to the problem of landlessness.

One of the problems associated with tourism is its vulnerability to negative news about unrest and social instability. In the 1990s, for example, reports of an outbreak of cholera scared off Japanese tourists in large numbers, which was a disaster for the hotels that had made efforts to target and cater to the Japanese market. The recovery in visitor arrivals was slow and it took a great deal of effort, including sending a cultural mission to Japan, to win back the confidence of this market. Other instances of negative reporting, such as the Gulf War and SARS, have led to a decline in visitor arrivals, the most severe occurrence coming after the bombings of 2002 and 2005. The downturn was exacerbated by news of the exodus of holidaymakers leaving the island, which makes a dramatic television story. The decrease had a direct impact on the incomes of those working in tourism with many being laid off as the employers ran out of money to pay them. Shortly after the 2005 bombings the Balinese-owned international airline company Air Paradise International became bankrupt, laying off more than 350 of their Balinese staff. Hotels suffered badly with occupation rates hovering between 10–30% during the three months after the bombings (*Bali Post*, December 19, 2005). The situation worsened with the revelation of a video containing the confessions of suicide bombers and threats from their leader, Noordin M Top, to continue attacking America and its allies.

The island was clearly in dire straits and there were intense debates about what could be done to restore Bali's economic resilience. After the bombings 2002 there were extensive discussions among government officers, academics and the members of NGOs via the media and various seminars about the need to re-focus on Bali's agricultural sector, which had been marginalized for decades in favour of tourism. Many argued this was the last chance for Bali to redevelop its agriculture, and might possibly be too late (MacRae 2005). The government allocated considerable amounts of money to farmers through their irrigation organizations known as *subak* and customary villages were also offered funding to encourage villagers to return to agriculture. When talking about agriculture, they meant agriculture in its broadest sense and not just paddy rice farming, but also plantations and the like.

Bali has a strong tradition of making ritual offerings and thus there is a demand for young coconut leaves (*janur*, *slèpan*) and fruit from which they are made. Local farmers are unable to keep up with the demand and thus supplies of fruit and leaves have to be imported from East Java and bananas from Kalimantan. Fruits imported from abroad are also used in offerings and such is the demand that apples, oranges, kiwi fruits, and mangos have to be imported from Australia and New Zealand. One outcome of these discussions was that the Balinese should intensify their fruit farming to meet local demand and reduce dependence on imports both from within Indonesia and abroad, and one of the local television stations, Bali TV, became involved in promoting the need for Bali to re-develop its agricultural sector. The Balinese mainly watch this television station and between February–April 2006 it ran a non-commercial advertisement to promote the use of local fruit for offerings and stop using imports. This was not an expression of anti-Western sentiment, but an urgent call for people to return to agriculture as tourism remained in the doldrums. Some members of the intelligentsia were more pessimistic, pointing out that agricultural products were generally low priced and that the irrigation system was in poor shape, and they argued that the encouragement of agriculture could only be part of the solution.

The Bali Post Media Group (BPMG) who also owned Bali TV was strongly in favour of these economic measures and launched a movement dedicated to Balinese values called *Ajeg Bali* (Bali Standing Strong). The term *Ajeg Bali* was formally introduced to the public for the first time in May 2002, during the launching of Bali TV, the then new local television network. The establishment of local television was made possible by a broadcasting regulation issued in 2002, which was part of Indonesia's decentralization process. During the era of centralized government, which ended with the fall of President Suharto in 1998, all television stations had to be based in the state capital of Jakarta, whereas the new regulation promotes the establishment of regionally-based television.

The introduction of local television helped to democratize information since news no longer went simply from the centre to the periphery but could originate locally. The Bali Post Group, moreover, seeks a wider role and aims to promote local culture as the foundation of Balinese identity. With decentralization regional identity came to be seen as more significant and thus the introduction of *Ajeg Bali* came to be seen as an important means of safeguarding and rejuvenating Balinese culture amid the growing phenomena of modernization and globalization. Bali TV placed billboards promoting *Ajeg Bali* around Denpasar city containing exhortations to use the Balinese language. For many Balinese this invitation had already become clichéd because the encouragement to preserve Balinese had already been exhaustedly promoted through various channels. *Ajeg Bali* largely remained a meaningless cultural expression, but this changed rapidly after the bombings of October 2002. This cultural expression then became a social and even political slogan, and it was popularly held that it was not only meant to be an invitation to preserve Bali's cultural but also a safeguard for the island, especially from terrorists. The expression *Ajeg Bali* thus became a powerful term to invite people to be highly vigilant within the context of security. It was widely held that without guaranteed security, tourists would not return and as happened in the Asian Crisis several banners were hung in public in 2002 announcing that 'Bali is safe, the tourists come'. Suddenly safety and security were given a higher priority than culture, since without tourists Bali would suffer and would no longer be *Ajeg* (standing strong), and thus the promotion of tourism was incorporated into the movement to attain *Ajeg* Bali. The desire to build an upstanding Bali became the rationale for many business and social activities, such as cleansing ceremonies, post-bombing orchestral performances, as well as Bali's clean-up days and sporting competitions.

As the tourism industry slowly picked up several months after the terrorist attacks, the already popular term *Ajeg Bali* became a buzzword, but started to lose its meaning. The public started to ask what its exact meaning was and questioned whether *Ajeg Bali* was cultural propaganda, a political movement, or even a marketing strategy devised by the media group that introduced and used it. A popular alternative suggestion was that it was an attempt to bring Bali back to its old past. Another interpretation was it was an attempt to '...safeguard Bali to make it forever strong', but there was also a fear that the ideas behind Ajeg Bali were essentially conservative and would end up making Bali stagnant. It had become a slogan that was neither easy to define nor suitable for everybody.

There is no single meaning of the term, but this has not stopped ordinary people and politicians from using it for their own means. In a political rally of 2004 prior to the most democratic Indonesian general election to date, the term *Ajeg Bali* was used frequently. Almost every presidential candidate to visit Bali accepted and used the term in their political rallies. They came to the Bali TV headquarters, either voluntarily or having been invited by the media group, not only be interviewed live but to also sign an inscription of *Ajeg Bali* at the station. Anyone who makes a visit to Bali TV will see dozens of signatures supporting Ajeg Bali attached to the wall of the station building, including those of former President Megawati Sukarnoputri and the current President Susilo Bambang Yudhoyono. The inscription that is signed by Megawati dated 2nd July 2004 says 'let us safe guard Bali! Keep the unity of the nation with the spirit of unity in diversity!' The other inscriptions have similar exhortations and through these the slogan *Ajeg Bali* came to national attention and there have been echoes in other local regions.

One of the substantive contributions of the Ajeg Bali movement is the setting up of a financial institution called *Koperasi Krama Bali* (Balinese Cooperative Community) in May 2005. Within seven months this nonprofit organization had gained more then 7,000 members from throughout Bali, who invested in the movement and the amount of money collected by May 2006 was Rp 5.1 billion (*Bali Post*, March 27, 2006). The money was invested to enable its members to borrow to set up productive ventures, notably small businesses, and its members sought opportunities in informal sector. They held several short courses on making *bakso* (meatball) and chicken soup and even went on to create also a Bali Donut. Significantly, it was the sector once dominated by migrants, but one change was the introduction of pork-balls, which the Balinese as Hindus could eat. The cooperative movement also runs short courses on wedding makeup, hairdressing and photography all of which aimed at providing jobless locals the opportunity to become self-employed. When the government tried to promote self-employment the response was unenthusiastic, but when the Bali Post Media Group took up the initiative, the response was overwhelming, indicating the power of the media in development in contemporary Indonesia. Another reason for the uptake was the severity of the economic downturn occasioned by the bombings and by 2006 many of the meatball outlets that had come into being were being run by people who had been trained by the cooperative movement's short course with some receiving loans. Many people who previously would not have looked for opportunities in the informal sector began to admire its resilience to threats.

One of the outcomes of the bombings has been the encouragement of self-employment, often with a distinctly Balinese flavour. Meatball sellers have, for example, installed shrines in their outlets and wear the distinctive Balinese headscarf, *udeng*, to work. These expressions of identity make sense in marketing since most of their buyers are locals. To emphasize their connection with their customers they address them in Balinese as opposed to Indonesian, in contrast to meatball sellers from elsewhere in the archipelago.

Exhortations to encourage the Balinese to work harder have also been couched in Hindu terms through references to the principle of *karma yoga*, which holds that people have a destiny to be in this world and that in addition to their obligations to perform rituals, they have a religious obligation to work; the work-shy thus have no right to live. It is also a practical outlook since by working people will be more likely to meet their needs and thus will be less likely to engage in criminal activities, which is strongly condemned in Balinese culture. An ambitious long-term goal of the Balinese Cooperative Community is to accumulate capital to buy shares in Bali's airport in the event of it being privatized, as well as in the Nusa Dua resorts when the hotel leases expire. These goals reflect the view that the Balinese should acquire a significantly larger stake than they do currently in businesses being operated in their own territory. It is still too early to judge the success of the movement with its ambition to simultaneously boost the economy and increase cultural pride.

Security Solution

Before the bombings in 2002, the Balinese paid little attention to security matters and were confident that peaceful conditions would prevail. They believed that the Gods must have saved their island in return for the frequency of the rituals conducted by the islanders. This attitude was taken for granted and was strengthened through the tourism discourse that Bali was safe and secure. In the early 1980s, a bomb blast occurred in a bus that was intending to cross Bali by ferry from Java and it was believed that a radical Muslim group that aimed to attack Bali had masterminded the attack. This incident intensified the Balinese belief that the Gods secured their island and that supernatural powers would prevent criminals and others who would harm Bali from entering the island. This belief, however, proved to be meaningless the instant that Amarozi and his colleagues discharged their explosives. The Balinese refrained from retaliating, aside from a few scuffles, partly in response to calls by government and opinion formers for restraint, but also because of the growing local realization that they had become caught up in a global struggle; the Security and Defense Minister at that time Matori Abdul Djalil had rapidly linked the bombings to the international terrorist network, al-Qaeda.

Security became a matter of urgency as government officers and other public figures and the representatives of all religious faiths in Indonesia gathered in the governor's house in Denpasar on the day after the bombings to call for an intensification of neighborhood security to guard against either violence or retaliation or another potential terrorist attack. The village security forces, *pecalang*, were mobilized to guard their own immediate areas and guesthouses were checked to try to identify the perpetrators and their network members. The increase in village security was one of the important security measures to be undertaken in the aftermath of the bombings and in the business sector Bali's police force met with members of the Bali Hotel Restaurant Association to set up an accreditation system for hotels. In addition to creating a security system for the starred hotels in Bali that looked like becoming terrorist targets as they had elsewhere in Indonesia, they set up a security accreditation system that aimed to help police more familiar with the layout of hotel buildings in case of kidnaps and terrorist threats. This accreditation system is carried out annually to ensure the security standards are maintained. So far, only star rated hotels have joined the accreditation system because only they have sufficient resources to participate. Since the 2002 bombings, most star-rated hotels have performed security checks on cars and guests entering their premises and these measures have been tightened since the attack on the Marriot Hotel in Jakarta in 2003.

The bombings of October 2005 revealed that territorial security was still insufficient since the bombers entered Bali via Java and went straight away to the targets in Kuta Square with its bars and cafes, and Jimbaran with its seafood restaurants that are popular with Westerners and Indonesians alike. The bombers did not show up on the security surveillance system that had been established, even though the police claimed that they had been working hard and were vigilant. It was clear that their resources and capacities were still insufficient to keep Bali entirely safe from terrorists. The Bali police chief, General Made Mangku Pastika, at that time, introduced Bali Security Council, which works to help police in fulfilling their responsibilities. Initially, this idea was rejected by some law makers who argued that the responsibility for security was to remain in the hands of police and could not be handed over to a new institution like the Bali Security Council. After several discussions and hearings between the Bali police chief and the Bali executive and legislative officers, the idea of the Bali Security Council was agreed, but its name, which could easily be associated with the United Nation 'Security Council' was changed into the Coordinating Body for Security in the Bali Region (*Badan Koordinasi Pengamanan Daerah Bali*).

The task of this institution is to collect various resources to help the police in performing their task in security matters. This institution could raise funds, receive donations from third parties, and liaise with police in other countries in working together in the field of security. Principally, the maintenance of security remains in the hands of the police. Since the police are discouraged from actively seeking money, donations received therefore avoid the appearance of corruption, and it is the task of the Coordinating Institution to provide resources and schemes that support the maximum performance of the police in guarding Bali.

Another security solution initiated by the Bali police is the formation of an honorary police force. Its members come from 27 ethnic and religious background, professional, and public figure, including representatives of the Indonesian Islamic Council (MUI) Bali branch, Forum of Religious Adherents, Christian Church, ethnics of Kalimantan, Celebes, Yogyakarta, Sumatra, and Social Club of Indonesian

Ethnic Tionghoa (Chinese). There are also members from Japan, Belgium, and India. Its membership will also include representatives from all consulates in Bali (*Tempointeraktif*, February 16th, 2006). The membership is limited to up to 200 people. They also support the program of Coordinating Institution Bali Security. Members of the honorary police have been trained in matters of security, channelling information to the police network. General Mangku Pastika said that the importance of the honorary police was, among other things, to build bridges to the various ethnic groups within society, as well as to lighten the police load in revealing cases that involve people from outside Bali (*Kompas*, December 18th, 2005). Since members of the honorary police come from diverse backgrounds, the institution could become a forum for exchanging ideas and consolidating attempts to maintain Bali's security for the common good. A representative from the Social Club of Indonesian Ethnic Tionghoa, Budi Argawa, for example, not only welcomed the establishment of the honorary police, but further said that

> Kami merasa ini sebagai upaya pemersatu berbagai golongan, organisasi dari berbagai daerah. Kami tidak keberatan membantu polisi dalam berbagai kesempatan. Justru dengan pembentukan polisi kehormatan ini merasa eksistensi kami lebih diakui (*Kompas*, December 18th, 2005).

> We feel this is an attempt to unite various groups and organizations from various regions. We do not mind helping the police on various occasions. In fact, with the establishment of the honorary police we feel that our existence is more acknowledged (*Kompas*, December 18th, 2005).

They are aware that in this global era threats could come from diverse sources and in order to combat them it would be necessary for members of these diverse communities to work hand in hand with the police.

Socio-political Solution

A 'Bloody Sunday' occurred in Singaraja, North Bali, on Sunday, October 26th, 2003. Two brothers, supporters of Golkar party, were killed in a clash with supporters of the ruling PDIP. Isolated clashes also occurred in several places around Bali as supporters of Megawati Sukarnoputri sought revenge for the perceived injustices meted out to them in the past, by members of Golkar, the party of former president, Suharto. The tension between these two groups escalated as each of them began to display their banners and party symbols alongside the island's streets in the run up to the country's first direct elections. The introduction of general elections is part of the radical changes taking place within Indonesia as the country attempts to transform itself into a more democratic nation. The public welcomed the reforms enthusiastically since they were freely able to express their political aspirations. The foreign media, however, closely scrutinized the transformation of the country into a democratic nation, and it is not surprising that the incident in Singaraja received widespread international coverage. The politically motivated killings in North Bali immediately became newsworthy both nationally and internationally, and the incident showed once again that Bali was still widely treated as a barometer of national stability. As the news was beamed around the world, the incident was treated as a potential threat to the country's process of democratization. Aware of this threat, the national leaders of the political parties and other observers urged the government to tighten the security apparatus to prevent such an incident from breaking out again and to defuse politically motivated conflict.

In Bali, the 'Bloody Sunday' incident was condemned, not just by political activists and government, but just about everybody, especially those working in the tourism sector. There were worries that such an incident could damage the peaceful image of the island at an especially sensitive time, as it struggled to recover from the first round of bombings and had just started to get a positive response. There was popular support for prioritizing the old slogan 'Bali Aman, Turis Datang' (Bali is safe, the tourists arrive) and it was widely felt that there was no point damaging the island's hard won image for short term political interests. As had occurred after the first bombings, the influential stakeholders – the leaders of political parties, the intelligentsia and religious figures – came together to consider the various socio-political solutions. They opted to implement a programme of community empowerment to combat strife and encourage peaceful political expression. They stressed traditional values and extolled the virtues of local genius through a combination of seminars, talkback radio and interactive television. Representatives of customary village councils and Hindu organizations undertook several tours of the island to encourage the Balinese people to strengthen their sense of brotherhood. Political activities as such were not banned, but those who wanted to participate were encouraged to keep a sense of perspective and to remain in harmony with their social environment. Traditional bodies and religious groupings were also explicitly kept out of any involvement in campaigning so that they would not become sullied by the political process and lose the respect of the islanders.

Latent tensions in Bali between the supporters of particular parties and clan groupings made the public worry that tensions might rise to the surface within this new and relatively open environment, and cause fierce clashes. Despite its generally positive image, historically, Bali has never been entirely free of various forms of conflict and violence (Robinson 1995). In 1965/66, for example, the Balinese were drawn into the mass killings of alleged Communist party members and fellow ideologues and in the early 1970s, the early period of New Order rule, many isolated incidents occurred in the north and south of the island as the government enforced acceptance of Suharto's military-backed party, Golkar. The imposition of Golkar's will, a process known locally as *kuningisasi*, led to conflict between the president's supporters and those still loyal to Sukarno and his nationalists. In the shadow of tourism, however, conflict was to be avoided at all costs. The 'Bloody Sunday' incident in North Bali had its roots in the start of Golkarization in 1971 and the long suppression that followed. Understandable though this pent up frustration was, the public were more concerned about the potential damage to Bali's image, which was just starting to recover in the aftermath of the 2002 crisis.

All attempts to promote peace in Bali within the context of the political process bore fruit. The general elections and the elections for the President and Vice President, as well as those of local office bearers, passed off smoothly in 2004 and 2005, and were free of conflict. Maintenance of stability was seen as being inseparable from the promotion of regional culture, and to reinforce this point local people were encouraged to wear Balinese attire when going out to cast their votes. In the final outcome, Bali's conflict-free elections contributed to the generally successful national picture, an important consideration given the media spotlight on the island.

What is significant about many of these attempts to find solutions to problems, and their implications, is the constant invocation of the sanctity of Balinese culture and the emphasis on the primacy of the needs of the community. These solutions neither come from the government in a top-down fashion nor totally from down below in bottom-up model, but from a dialogue involving many leading stakeholders. These interested parties have proved to be quite cohesive and able to act quickly in emergencies, such as that occasioned by the Bali bombings. In particular, the inter-faith forum proved to be effective in preventing inter-ethnic conflict which threatened to break out in the aftermath of the Bali bombings of 2002 (see Chapter 9). If there had not been terrorist attacks then it seems likely that the economy would have been in reasonable shape, and there would not have been a need to explore new forms of employment, such as meatball selling. The bombings clearly acted as a catalyst in the formation the Balinese Cooperative Community, but they were not the only reason why this body came into existence. Despite the introduction of emergency measures to cope with the failing economy, the islanders remain undeterred from developing their economy through tourism, notwithstanding the ongoing threat posed by terrorism. The widespread feeling that Bali has something special to offer in cultural terms, to both attract foreign visitors and to cope with global threats, has buoyed the islanders' resilience.

Chapter 13

Conclusion

Every culture has its talents and some become known worldwide for one in particular, to the extent that the name of the culture becomes inexorably linked with it. In their generous and friendly reception of guests and strangers, the Balinese have a talent for creating and reproducing beautiful experiences that go beyond what would normally be called hospitality. The style in which visitors are received has been imitated around the world and it is no surprise that there exist Bali-style beach houses in places as far away as the Maldives and Bali-themed restaurants in distant London and elsewhere. Designers around the world recognise the genre and thus there are Bali clothing and jewellery fashions, Bali-style gardens, Bali-type shop interiors and Bali-inspired songs, film, and theatre performances and so on. Given the relative cheapness of Balinese export handicrafts in relation to quality, it is quite easy to create a distinctly Balinese atmosphere in far-off lands. Bali is by no means unique in this achievement and one has only think of how other cultures – for example, Thai, Mexican, Italian and Moroccan – are widely associated with a particular style of living and above all, eating. Bali is, however, a relatively small culture in global terms and it would not be unreasonable to suggest that it is punching above its weight.

Despite the crises that have engulfed the island in recent years, the name Bali still retains many of its pleasurable connotations, and its popularity, as a brand name shows no sign of waning. For some visitors, however, the island has become unacceptably crowded and urbanised with hotels, shops and malls jostling for space. But for others the island has retained its charm, and remains an idyllic place for recreation, tranquillity and even spiritual fulfilment: the point is that Bali as a brand means many different things to many people. Bali has, for example, been selected twice to hold major international conferences on healing and this has enabled the island to renew, re-position and reinforce its emerging alternative image of tourism. Hardworking and increasingly health conscious visitors, especially Asians, are increasingly opting for designer and boutique hotels with fully equipped spas, and, though Bali is hardly alone in going down this route, it appears to be sufficiently capable of projecting its own distinctive variety.

Given that much of the literature on globalization seems to suggest that such particularities should cease to exist with the rapid movement across borders of capital, goods, labour, ideas and technologies, why should identities such as 'Bali' continue to burn so brightly? With regard to Bali there would appear to be three plausible reasons for this profile, all of which are closely interwoven and dynamically related to one another. First and above all, there is tourism, which is crucial to the island's development and long-term prosperity. But this industry brings with it powerful homogenising forces and in order to continue to prosper in this marketplace Bali needs to maintain a distinct identity. If Bali is to be recognised in a world seemingly replete with island paradises, not only abroad but close at hand within Indonesia, then it must offer something distinctive in order to attract the all important tourists, both internationally and domestically. This distinction has to have both mass appeal on the one hand and niche interest on the other, and Bali has thus far managed this by offering a kind of smorgasbord of resorts: Ubud for the upmarket and artsy, Sanur for lovers of a seaside urban village, Kuta for the clubbers and surfers, Seminyak for the cool and well-heeled.

Second, the Balinese are what anthropologists have loosely called 'traditionalists' in that they seek to preserve a way of life that has special meaning for them. In a sense all viable and ongoing cultures have this sense of traditionalism, but some make inordinate efforts to screen out what they see as harmful in the modern world and to preserve what they perceive to be special, even when it makes no apparent sense to the outside observer. In this respect the Balinese may be likened to a variety of cultures around the world that have contact with tourism, whether it be the Maasai, heirs to a distinctive African pastoral way of life, or the islanders of the Maldives, guardians of a distinctive Islamic civilisation. What is special about the Balinese, no matter how shrouded in mythology, is the sense that they are the modern exemplars of the old way of life of the Javanese kingdom of Majapahit, which was once central for much of present day Indonesia. The Balinese are not simply Majapahit's curators, but represent its living and breathing descendents, and for many Balinese this fabled kingdom has a contemporary reality and continues to exist to this day.

Third, there is Bali's position as a cultural and religious minority within the Indonesian nation state, a situation that is intensely scrutinised and adjusted by the island's intelligentsia in its dealings with the government in Jakarta. Parliamentary bills that appear to threaten the integrity of Bali's culture, such as the proposed pornography law in 2006, are opposed vehemently, even though Bali is a not a 'special region' like Aceh or Yogyakarta within Indonesia. The term 'freedom for Bali' is evoked when the going gets rough, but there is as yet no serious basis for an independence movement, and the expression is best regarded as form of rhetorical weaponry for lambasting the central authorities. The Balinese willingly erect statues in honour of the struggle for independence and when given a chance to vote they vote overwhelmingly for mainstream national parties such as Megawati's PDIP in 1999. Bali's economy is any case closely interwoven with that of the rest of Indonesia whether through investments from Jakarta or the growth in domestic tourism.

Bali's development processes and strategies have been primarily concerned with tourism for half a century, though the industry goes back much further, and this is reflected in the literature and the debates that have emerged. Somewhat unusually for a book on tourism, this study has examined works of realist literary representation: how the writings depict Balinese interactions with westerners, the motives of both groups, and Balinese people's perceptions of westerners; and how the writers deal with the significance of these interactions within the context of the Balinese social system. In Balinese cultural discourse in the 1990s globalization became a major theme, and many of the authors discussed in Chapter 6 attempted to come to terms with the process through the theme of interacting with foreigners. What this book does in particular is to examine the kinds of cultural awareness that the Balinese authors strove to create or reproduce in depicting the meeting of different societies. These works are widely read on the island and tap into popularly expressed concerns about the presence of foreigners who are clearly indispensable, but raise a series of worries. The fact is that foreigners take an interest in many things Balinese, often re-shuffling what existed before they arrived in ways that renders the term 'impacts of tourism' inadequate. These relationships with foreigners were important from the outset in shaping how Bali tourism would develop, the best known of these being that of the Russian-German artist Walter Spies and Cokorda Agung Sukawati of the court of Ubud in the inter-war years.

The way certain villages have become involved in tourism is also more complex than what is suggested by the term 'impact'. These settlements have often become conservation minded and proud of their role in interpreting what it means to be Balinese, though the intelligentsia with its scholarly concerns worries about authenticity and accuracy on one hand, while praising the islanders' cultural dynamism on the other. Awareness of tourism among these villagers also opens new economic opportunities that are not just locally bound, and the islanders are increasingly finding their way into the global workforce outside Bali, but not just in concerns that are directly related to tourism, such as the provision of hospitality on cruise ships, but also in a myriad of professions concerned with the creative sector, ranging from music and media to the visual arts and fashion.

Since the start of the 21st century the combination of the intensification of the process of democratization and the experience of terrorism have not only integrated Bali more closely into national debates, but have made the islanders more aware of the global context in which their economy operates and how their island is perceived abroad. The Balinese are thus compelled to think not only about the benefits of development, but also the accompanying problems, some of which, like the bombings, are very hard to predict and difficult to prevent when you are so globally exposed.

These processes of accelerated convergence has brought many opportunities and cheap goods to Bali, and one has only to look at the proliferation of mobile phones on the island to know that some of these have been heartily welcomed. There is, however, a downside and one has only to look at how the recent economic and political crises which have engulfed Indonesia, have had a major impact on Bali. The island was somewhat shielded from the Asian Crisis that tore apart the Indonesian economy, though this seems to have been due to special factors such as the need for the Jakarta elites to protect their investments and Bali's status as a refuge for persecuted Chinese. Its special status, however, has offered it no protection in the so-called War on Terror; in fact quite the opposite since Bali's renown as a tourist resort has made it a magnet for terrorists wanting to boost their profiles and to make their causes known internationally.

In the aftermath of the 2005 bombings the question that was on many people's lips in Bali was whether or not tourism would recover like it did after the horrific attacks of 2002. At the same time there were calls from the intelligentsia for the Balinese economy to diversify to mitigate the effects of global turbulence, some of them unrealistic such as the return to agriculture, but in many ways this was already happening with considerable successes in some sectors such as arts and handicrafts. The real estate sector also appeared to have been relatively unscathed by the bombings with demand not only from wealthy urban Indonesians, notably the Chinese, but also from expatriates from a great range of countries, many of whom had prior work experience in Indonesia, often choosing to stay on in Bali after retirement. The tourism sector was in any case starting to change with a greater emphasis on domestic tourism and visitors from the new tourist providing countries of Asia: Taiwan, Korea, Malaysia and China. The fall in prices that came after the 2002 bombings appears to have accelerated this change as the 'new tourists' took advantage of cut-price deals.

Some of the Balinese responses – such as the mobilization of village wards and village security – to the crises are probably culturally specific, but the Balinese model of crisis management – a constant reiteration of their desire for peace, joint prayers with other religious groupings, media restraint, and the rapid involvement of opinion leaders – may have wider applications in coping with the impacts of terrorism and conflict within the context of tourism. It is widely appreciated in Bali that if tourism is to be sustained, then security is a prerequisite. The bombings in 2002 and 2005, and ongoing threat of global terror have changed the Balinese's perceptions of the presence of security in hotel, tourism facilities, and shopping centres. Before the attack, the overt security was avoided because it was feared that it would engender feelings of insecurity, but after the attacks the presence of security became more apparent.

It is difficult to appreciate the effects of terror attacks on tourism purely in writing, and it was Richard Butler who suggested that the authors of this book should try to capture this by using the Tourism Area Life Cycle approach. The data was not by any means perfect, but the ensuing graph did help to visualize Bali's tourism trajectory and its responses to crises. We also tentatively suggested that the graph could help in plotting the recovery, though the island was struck by a second attack shortly after Bali TALC was produced and at the time of publication it was too early to see what the fallout would be.

Since the arrival of the first Dutch ships in the 16th century and Bali's piecemeal absorption into the empire of the Netherlands East Indies between the mid 19th and early 20th centuries, the island has been drawn into global networks of trade and cultural interchange. The islanders have had to cope with a succession of foreign traders, seaborne expeditions, colonial administrators, artists and academics, and since the early 20th century, tourists. With the exception of the death struggles of the royal households of Badung and Klungkung in 1906 and 1908 respectively, these foreigners have largely been accepted and even welcomed by the Balinese. The global significance of these arrivals was often not apparent to the islanders, but when global agencies came in a more concrete form, no matter how well intentioned, such as UNESCO in the case of the mother temple of Besakih, there was a strong reaction. But it was not UNESCO that the Balinese were objecting to, but how the powerful bureaucracy in Jakarta, which the Balinese at that time had a good reason to be suspicious of, would interpret this organisation's recommendations.

Angry responses also surfaced during the nationally lead tourism campaign that followed the 2002 bombings with its slogan 'Bali for the world'. Many felt that this was a step too far, a supine reaction to globalization, because Bali should be for

Bali, as opposed to surrender and exploitation on behalf of the world. At the time it was suggested that the slogan should be reversed to become 'the World for Bali', though its chauvinistic associations appears to have been lost on the new slogan's advocates.

Likewise, when the island was drawn into the imaginary clash of civilizations through the activities of the bombers, the Balinese rose up defiantly but not to punish the terrorists' co-religionists but to express their support for peace and inter faith dialogue. The interfaith group gathered for joint prayers and it was agreed that these would be held annually on the anniversary of the bombings. The Peace Monument close to the site of the bombings in Legian bears the names of victims from so many countries, and has become a local icon with a truly international dimension. It is a sad irony that another important local monument that concerns events of global significance that sent the name of Bali reverberating worldwide almost a hundred years earlier concerns a tragic occasion, the *puputan* in Badung.

The islanders clearly had an economic imperative – the restoration of their lucrative tourism industry – but it is not the only reason, not least because not everybody who gave their support is involved in tourism. The rituals were not only an expression of religious belief but also a reaffirmation of the islanders' solidarity with the Indonesian nation and its commitment to remaining a multi faith nation as enshrined in the national philosophy of *Pancasila*, inaugurated by Sukarto, the first president, and supported and endorsed by the country's leaders ever since.

The Balinese are also keenly aware that their blood was also spilled in a horrific explosion alongside many Australians in an event that has become one of the modern definers of contemporary Antipodean nationhood. Given the often fractious nature of Indonesian-Australian border relations, the Balinese widely regard themselves as a kind of cultural bridge and have taken to their hearts the expressions of togetherness in the aftermath of the bombings, such as wreaths laid at the site of the former Sari Nightclub baring the motto 'Bali and Australia – Friends for Ever', a reference to the famous Australian graduation song, 'Friends for Ever'.

The processes that were documented in the widely read works of Vickers and Picard have been thrown into sharp focus in the early 21st century as globalization has been intensified through ongoing tourism development, the ubiquitous media, health scares and the threat of terrorism. But as just as these forces have shaped the island, local agencies and actors have arisen to nurture and develop Bali's local talent and identity, and to provide creative responses. It is not the case that the essence of Bali has survived uncorrupted into modern times, though the tourism industry and its media/marketing henchmen would largely like us to see it that way, but that the island's culture has survived by responding flexibly to stimuli from the exterior. Instead of the homogenization and erosion of local culture, there is the use of culture as a strategic resource to be remoulded and deployed not only in response to external pressures, but to be used entrepreneurially to secure the advantage. Balinese traditional values are fiercely conserved, but are not inflexible and are often modified after intense debate.

Bibliography

Aa, B.J.M. van der, Groote, P.D. and Huigen, P.P.P. (2005), 'World heritage as NIMBY? The case of the Dutch part of Wadden Sea', in D. Harrison and M. Hitchcock (eds), *The Politics of World Heritage: Negotiating Tourism and Conservation* (pp. 11–21) (Clevedon: Channel View Publications).

Adams, K. (1997), 'Touting touristic "primadonas": tourism, ethnicity, and national integration in Sulawesi', in M. Picard and R.E. Wood (eds), *Tourism, Ethnicity and the State in Asian and Pacific Societies* (pp. 155–80) (Honolulu: University of Hawaii Press).

Adams, M.J. (1973), 'Structural aspects of a village art', *American Anthropologist*, 71 (1), 265–79.

Aditjondro, G.J. (1995), 'Bali, Jakarta's Colony: the Domination of Jakarta-Based Conglomerates in Bali's Tourism Industry and its Disastrous Social and Ecological Impact' (Working paper No. 52 Murdoch Asia Research Centre Sept/Oct).

Agung, IAAG. (1989), *Bali Pada Abad XIX* (Bali in the 19th century) (Yogyakarta: Gajah Mada University Press).

Anderson, B. (1973), 'Notes on contemporary Indonesian political communication', *Indonesia*, 16: 155–80.

Anderson, B. (1983), *Imagined Communities: Reflections on the Origin and Spread of Nationalism* (London: Verso).

Aridus, IM. (2002), 'Bali loved by the world', *Bali Post* (online), 22 December 2002.

Aryantha Soethama, IG. (1988), Suzan. *Sarinah* (4 series, nos. 151–254, 4 July–28 August).

Aryantha Soethama, IG. (1995), *Sang Juara, Sembilan Desa Terpilih di Bali* (The Champion, Nine Selected Villages in Bali) (Denpasar: Biro Humos dan Protokol Setwilda).

Aziz, H. (1995), 'Understanding attacks on tourists in Egypt', *Tourism Management*, 16 (2), 91–95.

Backhaus, N. (1998), 'Globalisation and Marine Resource Use in Bali', in V.T. King (ed.), *Environmental Challenges in South-East Asia* (pp. 169–192) (Richmond: Curzon).

Badan Pusat Statistik (2001), *Bali dalam Angka* (Bali in Figures) 2001, (Denpasar: Badan Pusat Statistik Provinsi Bali).

Badung Tourism Promotion Board (1978), *The ABC of Bali: A Guide to the Island of Bali* (Sanur: P.T. Bap).

Bagus, I Gusti Ngurah (1996), 'The play "women's fidelity": literature and caste conflict in Bali', in A. Vickers (ed.), *Being Modern in Bali, Images and Change* (pp. 92–114) (Monograph 43/Yale Southeast Asian Studies).

Bali Tourism Statistic (1999), *Bali Tourism Statistics* 1999 (Denpasar: Bali Government Tourism Office).
Bali Tourism Statistic (2004), *Bali Tourism Statistics* 2004 (Denpasar: Bali Government Tourism Office).
Baraas, F. (1983), *Léak* (Evil spirit) (Jakarta: Balai Pustaka).
Barth, F. (1969), 'Introduction', in F. Barth (ed.), *Ethnic Groups and Boundaries: the Social Organisation of Cultural Difference* (pp. 9–38) (Oslo: Norwegian University Press).
Bateson, G. and M. Mead (1942), *Balinese Character: A Photographic Analysis* (New York: New York Academy of Sciences).
Baudrillard, J. (1983), *Simulations* (New York: Semiotext(e)).
Baum, V. (1937), *Tale from Bali* (Garden City: Doubleday).
Belo, J. (ed.) (1970), *Traditional Balinese Culture* (New York: Columbia University Press).
Bird, G.W. (1897), *Wanderings in Burma* (Bournemouth: F.J. Bright and Sons).
Boeke, J.H. (1953), *Economics and Economic Policy of Dual Societies, as exemplified by Indonesia* (New York: Institute of Pacific Relations).
Boon, J.A. (1977), *The Anthropological Romance of Bali 1597–1972; Dynamic Perspectives in Marriage and Caste, Politics and Religion* (Cambridge: Cambridge University Press).
Bras, K. and Dahles, H. (1999), 'Massage Miss? Women entrepreneurs and beach tourism in Bali', in K. Bras and H. Dahles (eds), *Tourism and Small Entrepreneurs: Development, National Policy, and Entrepreneurial Culture: Indonesian Cases* (pp. 35–51) (New York: Cognizant Communications Corporation).
Bras, K. and Dahles, H. (1999), 'Pathfinder, gigolo, and friend: diverging entrepreneurial strategies of tourist guides on two Indonesian islands', in H. Dahles and K. Bras (eds), *Tourism and Small Entrepreneurs: Development, National Policy and Entrepreneurial Culture: Indonesian Cases* (pp. 128–45) (New York: Cognizant Communication Corporation).
Butler, R.W. (1980), 'The concept of a tourist area cycle of evolution: implications for management of resources', *Canadian Geographer*, 24: 5–12.
Butler, R.W. (1993), 'Pre- and post-impact assessment of tourism development', in D.G. Pearce and R.W. Butler (eds), *Tourism research: critiques and challenges* (pp. 135–55) (London: Routledge).
Butler, R.W. (2000), 'The resort cycle two decades on', in B. Faulkner, G. Moscardo and E. Laws (eds), *Tourism in the 21st Century, lessons from experience* (pp. 284–99) (London: Continuum).
Butler, R.W. (2004), 'The tourism area life cycle in the twenty-first century', in A.A. Lew, C.M. Hall and A.M. Williams (eds), *A Companion to Tourism* (Blackwell: Oxford).
Butler, R.W. (2006) 'The concept of a tourist area cycle of evolution: implications for management of resources', in R.W. Butler (ed.), *The Tourism Area Life Cycle Vol. 1: Applications and Modifications* (pp. 3–12) (Clevedon: Channel View Publications).
Chard, C. (unpublished paper). 'Women who transmute into tourist attractions: spectator and spectacle on the Grand Tour'.

Cheater, A.P. (1995) 'Globalisation and the new technologies of knowing: Anthropological calculus or chaos?', in M. Strathearn (ed.) *Shifting Contexts: Transformations in Anthropological Knowledge* (London: Routledge).

Christie, J.W. (1993), 'Ikat to Batik? Epigraphic Data on Testiles in Java from the Ninth to the Fifteen Centuries', in M.L. Nabholz-Kartaschoff, R. Barnes and D.J. Stuart-Fox (eds), *Weaving Patterns of Life* (pp. 11–29). (Basel: Museum of Ethnography).

Chulov, M. (2003), 'Paddy's bar rises from the ashes', *The Australian*, 4 August.

Cleverdon, R. (1993), 'Global tourism trends: influences, determinants and directional flows', in D.E. Hawkins, J.R.B. Ritchie, F. Go and D. Frechtling (eds), *World Travel and Tourism Review Indicators* (pp. 81–89), Trends and Issues 3 (Wallingford: CAB International).

Clune, F. (1940) *To the Isles of Spice with Frank Clune...* (Sydney: Angus and Robertson).

Coast, J. (1966), 'Bali revisited', *Dance Magazine*, November 1966: 46–49, 72.

Cole, S. (2001), 'Appropriate tourism: megaliths and meaning in eastern Indonesia', *Indonesia and the Malay World*, 31: 89 (March), 140–150.

Colmey, J. and Liebhold, D. (1999), 'All in the family', *Time*, 24 May, 36–9.

Cork, Vern (translator) (1996), *Bali Behind the Seen* (Sydney: Darma Printing).

Couteau, J. (2003), 'After the Kuta bombing: in search of the Balinese soul', *Antropologi Indonesia*, 70, 41–59.

Covarrubias, M. (1937 [1956]), *Island of Bali* (New York: Alfred A. Knopf).

Crick, M. (1985), 'Tracing the anthropological self: quizzical reflections on field work, tourism and the ludic', *Social Analysis*, 17, 71–92.

Cukier, J. and Wall, G. (1994), 'Informal tourism employment: vendors in Bali, Indonesia', *Tourism Management*, 15(6), 464–67.

Cukier, J. (1996), 'Tourism employment in Bali: trands and implications', in R. Butler and T. Hinch (eds), *Tourism and Indigenous People* (pp. 49–75) (London: Thomson International Business Press).

Dahles, H. (1999), 'Tourism and small entrepreneurs in developing countries: a theoretical perspective', in H. Dahles and K. Bras (eds), *Tourism and Small Entrepreneurs: Development, national Policy and Entrepreneurial Culture: Indonesian Cases* (pp. 1–19) (New York: Cognizant Communication Corporation).

Dalton, B. (1978), *Indonesia Handbook* (2nd edition) (Vermont: Moon Publications).

Dann, G. (1996), 'The people of tourist brochures', in T. Selwyn (ed.), *The Tourist Image* (pp. 61–82) (Chichester: Wiley).

Dann, G. (1996a), *The Language of Tourism: A Sociolinguistic Perspective* (Wallingford: Cab).

Darma Putra, I.N. (1996), 'Pergeseran estetika dalam cerpen-cerpen Gde Aryantha Soethama (Aesthetic shifts in Gde Aryantha Soethama's short stories), *Bali Post*, 4 February, 10.

Darma Putra, I.N. (2003), 'A literary mirror, Balinese reflections on modernity and identity in the twentieth century' (Unpublished thesis, The University of Queensland).

Darma Putra, I.N. (2003a), *Penganugerahan Karya Karana Pariwisata 2003, Profil Penerima Penghargaan* (Denpasar: Dinas Pariwisata Bali).

Darma Putra I.N. (2006), 'The Bali bombs and the tourism development cycle', *Progress in Development Studies*, 6: 2, 157–166.

de Jonge, H. (2000), 'Trade and ethnicity: Street and beach sellers from Raas on Bali', *Pacific Tourism Review*, 4, 75–86.

de Kadt, E. (1979), *Tourism: Passport to Development* (Oxford: Oxford University Press).

De Zoete, B. and Spies, W. (1938), *Dance and Drama in Bali* (London: Faber and Faber).

Dewa Made, T. and Pakpahan, A. (1990), 'Towards and Agriculture that Supports Tourism', in "Proceedings, The International Seminar on Human Ecology, Tourism, and Sustainable Development" (pp. 188–93) (Denpasar: Bali-HESG).

Dinas Kebudayaan Bali (1999), 'Heritage site conservation Pura Agung Besakih' (the mother temple) *Report Bali urban infrastructure project*, 4 (Dinas Kebudayaan Provinsi Bali in association with SAGRIC Int. Pty. Ltd.).

Eagleton, T. (1986), *Literary Theory, An Introduction* (Oxford: Basil Blackwell).

Eco, Umberto (1986), *Faith in Fakes: Travels in Hyperreality* (London: Minerva).

Edgel, D.L. (1990), *International Tourism Policy* (New York: Van Nostrand Reinhold).

Eidheim, H. (1971), *Aspects of the Lappish Minority Situation* (Oslo: Norwegian University Press).

Eiseman, F.B. (1989), *Bali Sekala and Niskala: Essays on Religion, Ritual and Art* (vol. 1). (Berekely: Periplus).

Ellis, E. (2004), 'I've been to Bali too', *The Weekend Australian Magazine*, September, 4–5: 28–31.

Erawan, IN. (2003), 'Recovery Pembangunan Bali Pasca Bom Bali', in IGB Sudhyatmaka Sugeriwa (ed.), *Bom Bali* (pp. 264–67) (Denpasar: Biro Humas dan Protokol Setda Propinsi Bali).

Eriksen, T.H. (1991), 'The cultural contexts of ethnic differences', *Man*, 26: 1, 127–144.

Flierhaar, H. te (1941) 'De aanpassing van het inlandsch onderwijs op Bali aan de eigen sfeer', Koloniale Studiën, 25, 135–159.

Forshee, J. (1998), 'Sumba asli: fashioning culture along expanded exchange circles', *Indonesia and the Malay World*, 26 (75), 106–123.

Francis, G. (1982), *Njai Dasima* [Miss Dasima], in Pramudya Ananta Tur (ed.), *Tempo Doeloe* (Past Time) (pp. 225–47) (Jakarta: Hasta Mitra).

Furnivall, J.S. (1968), *Colonial Policy and Practice* (Cambridge: Cambridge University Press).

Ganjanapan, A. (2003), 'Globalization and the dynamics of culture in Thailand', in S. Yamashita and J.S. Eades (eds), *Globalization if Southeast Asia: Local, National and Transnational Perspectives* (pp. 126–41) (New York: Berghahn Books).

Geertz, C. (1961), 'Review of J.L. Swelengrebel et al., *Bali: Studies in life, thought and ritual*', *Bijdragen tot de Taal-, Land-en Volkenkunde*, 117: 498–502.

Geertz, C. (1963), 'The integrative revolution: primordial sentiments and civil politics in the new states', in C. Geertz (ed.), *Old Societies and New States: The Quest for Modernity in Asia* (pp. 105–57) (London: Free Press of Glencoe).

Geertz, C. (1973), *The Interpretation of Cultures* (New York: Basic Books).

Geertz, C. (2004), 'Review of Leo Howe's *Hinduism and hierarchy in Bali* and David J. Stuart-Fox's *Pura Besakih: temple, religion and society in Bali*', *L'Homme*, 169, 285–87.

Geertz, H. (1991), *State and Society in Bali* (Leiden: KITLV Press).

Gelebet, IN. (2002), 'Dari Raja Purana hingga warbunia', *Sarad*, 22 January, 42–43.

Geriya, I.W. (2003), 'The impact of tourism in three tourist villages in Bali', in S. Yamashita and J.S. Eades (eds), *Globalization if Southeast Asia: Local, National and Transnational Perspectives* (New York: Berghahn Books).

Giddens, A. (1991), *The Consequences of Modernity* (Cambridge: Polity Press).

Go, F. and Pine, R. (1995), *Globalization Strategy in the Hotel Industry* (London: Routledge).

Godley, M.R. (1999), 'The Chinese southern diaspora', *International Institute for Asian Studies Newsletter*, 19, June.

Goffman, E. (1958), *The Presentation of Self in Everyday Life* (Garden City, New York: Anchor).

Gombrich, E.H. (1969), *Art and Illusion* (Princeton: Princeton University Press).

Gomez, T.G. (1999), *Chinese Business in Malaysia: Accumulation, Ascendance, Accommodation* (Richmond: Curzon).

Goris, R. (n.d.) 'Godsdienst en Gebruiken in Bali. Observations on the Customs and Life of the Balinese', in *Bali* (Batavia: Travellers' Official Information Bureau for Netherkand India, 1931).

Graburn, N. (ed.) (1976), *Ethnic and Tourist Arts: Cultural Expressions From the Fourth World* (Berkeley: University of California Press).

Hall, C.M. (2000), 'Tourism in Indonesia: the end of the New Order', in C.M. Hall and S. Page (eds), *Tourism in South and South-East Asia* (pp. 157–66) (Oxford: Butterworth-Heinemann).

Hall, C.M. and O'Sullivan,V. (1996), 'Tourism, political stability and violence', in A. Pizam and Y. Mansfield (eds), *Tourism, crime and international security issues* (pp. 105–21) (Chichester: John Wiley).

Hall, C.M., Timothy, D.J. and Duval, D.T. (eds) (2003), *Safety and security in tourism: relationships, management and marketing* (New York: The Haworth Press).

Hampton, M. (1998), 'Backpacker tourism and economic development', *Annals of Tourism Research*, 25 (30), 639–60.

Hampton, M. (1998b), 'Backpacker tourism and economic development in Yogyakarta' (unpublished conference paper, ASEASUK).

Hanna, W.A. (1976), *Bali Profile: People, Events, Circumstances (1001–1976)* (New York, American Universities Field Staff).

Harrison, D. (2003), 'Themes in Pacific island tourism', in D. Harrison, (ed.) *Pacific island tourism* (pp. 1–23) (New York: Cognizant Communication Corporation).

Heberer, G. and Lehmann, W. (1950), *Die Inland-Malaien von Lombok und Sumbawa* (Gőttingen: Gőttingen University Press).

Heers, W. (1948), *An Anthropological Survey of the Eastern Little Sunda Islands, the Negrito's of the Eastern Little Sunda Islands, the Proto-Malay of the Netherlands East Indies* (Amsterdam: Koninklijk Vereeningen Indisch Instituut).

Henderson, J.C. (1999), 'Southeast Asian tourism and the financial crisis: Indonesia and Thailand compared', *Current Issues in Tourism*, 2: 4, 294–303.

Higham, J. (2000), 'Thailand: Prospects for a Tourism-led Economic Recovery', in C.M. Hall and S. Page (eds), *Tourism in South and South-East Asia* (pp. 129–143) (Oxford: Butterworth/Heineman).

Hilbery, R. (1983), *Tjokorde Gde Agung Sukawati: Ubud 1910–1978 Autobiography* (Denpasar: Mabhakti and Southeast Asian paper No. 14, University of Hawaii).

Hitchcock, M. (1995), 'Inter-ethnic relations and tourism in Bima-Sumbawa', *Sojourn*, 10(2), 233–58.

Hitchcock, M. (1996), *Islam and Identity in Eastern Indonesia* (Hull: Hull University Press).

Hitchcock, M. (1998), 'Tourism, Taman Mini and national identity', *Indonesia and the Malay World*, 26 (74) (June), 124–35.

Hitchcock, M. (2000), 'Ethnicity and tourism entrepreneurship in Java and Bali', *Current Issues in Tourism* 3 (3), 204–25.

Hitchcock, M. (2001), 'Tourism and total crisis in Indonesia: The case of Bali', *Asia Pacific Business Review*, 8 (2), 101–120.

Hitchcock, M. (2002), 'Zanzibar Stone Town joins the imagined community of World Heritage Sites', *International Journal of Heritage Studies*, 8(2), 153–66.

Hitchcock, M. (2004), 'Margaret Mead and tourism: anthropological heritage in the aftermath of the Bali bombings', *Anthropology Today*, 20 (3): 9–14.

Hitchcock, M. and Darma Putra, IN. (2004), 'Bali bombings: tourism crisis management and conflict avoidance', *Current Issues in Tourism*, 8 (1), 62–76.

Hitchcock, M. and Norris, L. (1995), *Bali: The Imaginary Museum* (Kuala Lumpur: Oxford University Press).

Hitchcock, M., King, V.T. and Parnwell, M.J.G. (1993), 'Introduction', in Hitchcock, M., V.T. King and M.J.G. Parnwell (eds) (1993), *Tourism in South-East Asia* (pp. 1–31) (London: Routledge).

Hobart, A., Ramseyer, U. and Leeman, A. (1996), *The Peoples of Bali* (Oxford: Blackwell).

Hobart, M. (ed.) (1993), *An Anthropological Critique of Development: The Growth of Ignorance* (London: Routledge).

Hobsbawm, E. and Ranger, T. (eds) (1983), *The Invention of Tradition* (Cambridge: Cambridge University Press).

Hoerip, S. (ed.) (1986), *Cerita Pendek Indonesia IV* (Indonesian Short Stories IV) (Jakarta: Gramedia).

Holt, C. (1967), *Art in Indonesia: Continuities and Change* (Ithaca, New York: Cornell University Press).

Holtzappel, C. (1996), 'Nationalism and Cultural Identity', in M. Hitchcock and V.T. King (eds), *Images of Malay-Indonesian Identity* (pp. 63–107) (Kuala Lumpur: Oxford University Press).

Houellebecq, M. (2002), *Platform* (London: Heinemann).

Howard, J. (1984), *Margaret Mead: A Life* (New York: Simon and Schuster).

Hubinger, V. (1992), 'The creation of Indonesian national identity', *Prague Occasional Papers in Ethnology*, 1, 1–35.

Hudson, K. (1987), *Museums of Influence* (Cambridge: Cambridge University Press).

Hughes-Freeland, F. (1993), 'Packaging dreams: Javanese perceptions of tourism and performance', in M. Hitchcock, V.T. King and M.J.G. Parnwell (eds), *Tourism in South-East Asia* (pp. 138–54) (London: Routledge).

Hulsius, L. (ed.) (1598), *Eerste Schiffart an die Orientalische Indien, so die Hollandisch Schiff im Mario 1597 Aussgafahren, und Augusto 1598* Weiderkommen, Versicht. Nuremburg.

Hutt, M. (1996), 'Looking for Shangri-La: from Hilton to Lamichane', in T. Selwyn (ed.), *The Tourist Image: Myths and Myth Making in Tourism* (pp. 49–60) (Chichester: John Wiley and Sons).

Jenkins, R. and Catra, N. (2004), 'Answering terror with art: Shakespeare and the Balinese response to the bombings of October 12, 2002' (unpublished paper).

Johnson, J. and Snepenger, D. (1993), 'Application of the tourism life cycle concept in the Greater Yellowstone Region', *Society and Natural Resources*, 6, 127–148.

Johnston, C.S. (2001), ' Shoring the doundations of the destination life cycle model, part 1: ontological and epistemological considerations', *Tourism Geographies*, 3 (1), 2–28.

Just, P. (2001), *Dou Donggo Justice: Conflict and Morality in an Indonesian Society*, (Lanham, Bolder, New York: Rowman and Littlefield Publishers).

Kadir Din (1992), 'The involvement stage in the evolution of a tourist destination', *Tourism Recreational Research*, 17 (1), 10–20.

Kagami, H. (2003), 'How to Live a Local Life: Balinese Responses to National Integration in Contemporary Indonesia', in S. Yamashita and J.S. Eades (eds), *Globalization in Southeast Asia: Local, National and Transnational Perspectives* (pp. 65–80) (New York: Berghahn Books).

Keers, W. (1948), *An Anthropological Survey of the Eastern Little Sunda Islands, the Negrito's of the Eastern Little Sunda Islands, the Proto-Malay of the Netherlands East Indies* (Amsterdam: Koninklijk Vereeningen Indisch Instituut).

Kerepun, M.K. (2002), 'Hindu dan Pura Besakih', *Sarad*, 22 January, 44–45.

Koke, L.G. (1987), *Our Hotel in Bali* (Wellington: January Books).

Korn, V.E. (1933), *De Dorpsrepubliek Tnganan Pagringsingan* (Santpoort).

Krause, G. (1988), *Bali 1912* (Wellington: January Books).

Lagiewski, R.M. (2006), 'The application of the TALC model: a literature survey', in R.W. Butler (ed.), *The Tourism Area Life Cycle Vol. 1: Applications and Modifications* (pp. 27–50) (Clevedon: Channel View Publications).

Lanfant, M-F. (1995) 'Introduction', in M-F. Lanfant, J.B. Allacock and E.M. Bruner (eds), *International Tourism: Identity and Change* (pp. 1–23) (London: Sage Publications Ltd).

Liefrinck, F.A. (1927) *Bali en Lombok* (Amsterdam).

Lindsey, T. (1997), *The Romance of K'tut Tantri and Indonesia* (Kuala Lumpur: Oxford University Press).

Lingard, J. (translator) (1995), *Diverse Lives. Contemporary Stories from Indonesia* (Kuala Lumpur: Oxford in Asia).

Guibernau, M. and Rex, J. (1997). *The Ethnicity Reader: Nationalism, Multiculturalism and Migration* (Cambridge: Polity Press).

Mabbett, H. (1987), *In Praise of Kuta – From Slave Port to Fishing Village to the Most Popular Resort in Bali* (Wellington: January Books).

Maccannell, D. (1992), *Empty Meeting Grounds: The Tourist Papers* (London: Routledge).

Macleod, D.V.L. (2004), *Tourism, Globalisation and Cultural Change: An island Community Perspective* (Clevedon: Channel View publications).

MacRae, G. (1992), 'Tourism and Balinese culture' (M.Phil dissertation in anthropology, University of Auckland).

MacRae, G. (1999), 'Acting global, thinking local in a Balinese tourist town', in R. Rubinstein and L.H. Connor (eds), *Staying Local in the Global Village: Bali in the Twentieth Century* (pp. 123–54) (Honolulu, University of Hawaii Press).

MacRae, G. (2005), 'Growing rice after the bomb: where is Balinese agriculture going?', *Critical Asian Studies*, 37 (2), June.

Magnis-Suseno, F. (1997), *Javanese Ethics and World View: The Javanese Idea of the Good Life* (Jakarta: P.T. Gramedia).

Maier, H. (1993), 'From heteroglossia to polyglossia: the creation of Malay and Dutch in the Indies', *Indonesia*, October, 56: 37–56.

Manda, I.N. (1978), 'Togog' [Statue], *Bali Post*, 7–15 November, 4.

Mann, R.I. (1994), *The Culture of Business in Indonesia* (Toronto: Gateway Books).

May, B. (1978), *The Indonesian Tragedy* (London: Routledge and Kegan Paul).

McKean, P.F. (1971), 'Pengaruh-pengaruh asing terhadap kebudayaan Bali: hubungan "hippies" dan "pemuda international'dengan masyarakat Bali masa Kini"', in I.G.N.B. (ed.), *Bali dalam Sentuhan Pariwisata*, pp. 21–27 (Denpasar: Fakultas Sastra Unud).

McLuhan, M. (1968), *War and Peace in the Global Village* (New York: McGraw-Hill).

Mead, M. and Bateson, G. (1942), *Balinese Character: A Photographic Analysis* (New York: Academy of Sciences).

Menkhoff, T. and C.E. Labig. (1996), 'Trading networks of Chinese entrepreneurs in Singapore', *Sojourn*, 11(1), 128–51.

Michaud, J. (1995), 'Questions about fieldwork methodology', *Annals of Tourism Research*, 22, 681–87.

Mohamad, G. (1994), 'Sekedarnya tentang Putu Wijaya' (A sight about Putu Wijaya), in Putu Wijaya short story collection *BLOK* (pp. xi–xv) (Jakarta: Pustaka Firdaus).

Nugroho-Heins, M.I. (1995), 'Regional Culture and National Identity: Javanese Influence on the Development of National Indonesian Culture' (unpublished paper EUROSEAS, Leiden).

Parnwell, M.J.G. (1993), 'Tourism and rural handicrafts in Thailand', in M. Hitchcock, V.T. King and M.J.G. Parnwell (eds), *Tourism in South-East Asia* (pp. 234–57) (London: Routledge).

Parsua, N. (1986), *Anak-Anak* (Children) (Jakarta: Balai Pustaka).

Pemberton, J. (1989), 'An Appearance of Order: A Politics of Culture in Colonial and Postcolonial Java' (PhD thesis Cornell University).

Persoon, G. (1986), 'Congelation in the melting pot, The Minangkabau in Jakarta', in P.J.M. Nas (ed.), *The Indonesian City: Studies in Urban Development and Planning* (Dordrecht: Foris Publications).

Peters, M. (1969), *International Tourism: The Economics and Development of International Tourist Trade* (London: Hutchinson).

Picard, M. (1993), 'Cultural tourism in Bali: national integration and regional differentiation', in M. Hitchcock, V.T. King and M.J.G. Parnwell (eds), *Tourism in South East Asia* (pp. 71–98) (London: Routledge).

Picard, M. (1995), 'Cultural heritage and tourist capital: cultural tourism in Bali', in M-F. Lanfont, J.B. Allcock and E.M. Bruner (eds), *International Tourism: Identity and Change* (pp. 44–66) (London: Sage).

Picard, M. (1996). *Bali: Cultural Tourism and Touristic Culture* (Singapore: Archipelago Press).

Picard, M. (1997), 'Cultural tourism, nation-building and regional culture: the making of Balinese identity', in M. Picard and R.E. Wood (eds), *Tourism, Ethnicity and the Sate in Asian and Pacific Societies* (pp. 181–214) (Honolulu: University of Hawaii Press).

Picard, M. (2003), 'Touristification and Balinization in a time of reformasi', *Indonesia and the Malay World*, 31(89), 108–18.

Picard, M. (1990), 'Kebalian orang Bali: tourism and the uses of "Balinese culture" in New Order Indonesia', *Rima*, 24: 1–38.

Piet, S.L. (1936), *Pengoendjoekan Poelo Bali Atawa Gids Bali* (Malang: The Paragon Press).

Pincus, J. and Ramli, R. (1998), 'Indonesia: from showcase to basket case', *Cambridge Journal of Economics*, vol. 22, pp. 723–34.

Pitana, I.G. (2002), *Apresiasi Kritis terhadap Kepariwisataan Bali* (Denpasar: Bali Tourism Board).

Pizam, A. and Mansfield, Y. (1996), 'Introduction', in A. Pizam and Y. Mansfield (eds), *Tourism, Crime and International Security Issues* (pp. 1–17) (Chichester: John Wiley and Sons).

Poyk, G. (1982), *Di Bawah Matahari Bali* (Under the Sun of Bali) (Jakarta: Sinar Harapan).

Prideaux, B. (1999), 'Tourism perspectives of the Asian financial crisis: lessons for the future', *Current Issues in Tourism*, vol. 2, no. 4, 279–293.

Rabasa, A.M. (2003), *Political Islam in Southeast Asia: Moderates, Radicals and Terrorists* (Oxford: Oxford University Press for the International Institute for Strategic Studies).

Raffles, T.S. (1817), *The History of Java* (2 vols) (London: Parbury and Allen, and John Murray).

Ramseyer, U. (1984), *Clothing, Ritual and Society in Tenganan Pegeringsingan* (Basel: Sonderabdruck aus der Verhandlungen der Naturforschenden Gesellschaft in Basel Band 95).

Rasta Sindhu, N. (1969), 'Sahabatku Hans Schmitter' (My Friend Hans Schmitter), *Horison*, July, 215–17.

Rata, I.B. (1997), 'The use of archeological remains in the development of cultural tourism', in W. Nuryanti (ed.), *Tourism and heritage management* (pp. 357–64) (Yogyakarta: Gadjah Mada University Press).

Reuter, T. (ed.) (2003), *Inequality, Crisis and Social Change in Indonesia, The Muted Worlds of Bali* (London: CurzonRoutledge).

Reuter, T.A. (2002), *Custodians of the Sacred Mountains: Culture and Society on the Highlands of Bali* (Honolulu: University of Hawaii Press).

Rex, J. (1986), *Race and Ethnicity* (Buckingham: Open University Press).

Rhodius, H. and Darling, J. (1980), *Walter Spies and Balinese Art* (Amsterdam: Terra Zutphen).

Richter, L.K. and Waugh, W.L. (1986), 'Terrorism and tourism as logical companions', *Tourism Management*, vol. 7, no. 4: 230–38.

Richter, L.K. (1994), 'The political dimensions of tourism', in B. Richie and C.R. Geoldener (eds), *Travel, Tourism and Hospitality Research* (New York: Wiley).

Rosemary, J. (1987), *Indigenous Enterprises in Kenya's Tourism Industry* (Geneva: UNESCO).

Rubinstein, R. and Connor, L.H. (1999), 'Introduction', in Rubinstein, R. and Connor, L.H. (eds), *Staying Local in the Global Village: Bali in the Twentieth Century* (pp. 1–14) (Honolulu, University of Hawaii Press).

Sabdono, D. and Danujaya, B. (1989), 'Kisah Cinta Seumur Visa' (Love as long as valid visa), in P. Kitley, R. Chauvel and D. Reeve (eds), *Australia di Mata Indonesia. Kumpulan Artikel Pers Indonesia 1973–1988* (Australia in the eyes of Indonesians. A collection of articles on Indonesian press, 1973–1988) (pp. 963–67) (Jakarta: Gramedia).

Said, E. (1978) *Orientalism* (New York: Pantheon).

Salah, W. (1996), 'Tourism and terrorism: sythesis of the problem with emphasis on Egypt', in A. Pizam and Y. Mansfield (eds), *Tourism, Crime and International Security Issues* (pp. 175–202) (Chichester: John Wiley and Sons).

Samudra, I. (2004), *Aku Melawan Teroris* (I Oppose Terrorism) (Solo: Jazera).

Sangger, A. (1988), 'Blessing of Blight? The Effects of Touristic Dance Drama on Village Life in Singapadu', in O. Lewin, and A.L. Kaeppler (eds), *Come Mek Me Hol' Yu Han': The Impact of Tourism on Traditional Music* (pp. 89–104) (Kingston: Jamaica Memory Bank).

Sanggra, M. (1975), *Katemu ring Tampaksiring* (Encounter in Tampaksiring) (Gianyar: Yayasan Dharma Budhaya).

Scholte, J.A. (1996), 'Identifying Indonesia', in M. Hitchcock and V.T. King (eds), *Images of Malay-Indonesian Identity* (pp. 21–44) (Kuala Lumpur: Oxford University Press).

Selwyn, T. (1996), 'Introduction', in T. Selwyn (ed.), *The Tourist Image: Myths and Myth Making in Tourism* (pp. 1–32) (Chichester: Wiley).

Shackley, M. (ed.) (1998), *Visitor management: case studies from world heritage sites* (London: Butterworth-Heinemann).

Shaw, B.J. and Shaw, G. (1999), 'Sun, sand and sales: enclave tourism and local entrepreneurship in Indonesia', *Current Issues in Tourism* 2(1): 68–81.

Sinclair, M.T. and and R. Vokes. (1993), 'The economics of tourism in Asia and the

Pacific', in M. Hitchcock, V.T. King and M.J.G. Parnwell (eds), *Tourism in South-East Asia* (pp. 200–213) (London: Routledge).

Smith, M. (2002), 'A critical evaluation of the global accolade: the significance of world heritage site status for maritime Greenwich', *International Journal of Heritage Studies*, 8(2): 137–52.

Smith, M.G. (1965), *The Plural Society in the British West Indies* (Berkeley and Los Angeles: University of California Press).

Spies, W. and De Zoete, B. (1938), *Dance and Drama in Bali* (London: Faber and Faber).

Stockwell, A.J. (1993), 'Early tourism in Malaya', in M. Hitchcock, V.T. King and M.J.G. Parnwell (eds), *Tourism in South-East Asia* (pp. 258–70) (London: Routledge).

Strapp, J.D. (1988), 'The resort cycle and second homes', *Annals of Tourism Research*, 15 (4): 504–16.

Stuart-Fox, D.J. (2002), *Pura Besakih: temple, religion and society in Bali* (Leiden: KITLV Press).

Suarsa, M. et al. (1992), 'I Made Sanggra, sebagai manusia dan pelopor penulisan puisi Bali modern serta tinjauan terhadap karya-karyanya' (I Made Sanggra, a pioneer in modern Balinese poetry with an analysis of his works) (Unpublished research manuscript, Udayana University).

Sudarmaji. (no date). *Different Styles of Painting in Bali* (Ubud: Neka Gallery).

Suharto (1975), 'Words of welcome', in *What and Who in Beautiful Indonesia* (Jakarta).

Sulistyawati, M.S. (2000), 'Tourism and geringsing textiles in Bali: a case study from Tenganan', in M. Hitchcock and Wiendu Nuryanti (eds), *Building on Batik: The Globalization of a Craft Community* (pp. 295–306) (Aldershot: Ashgate).

Sullivan, G. (1999), *Margaret Mead, Gregory Bateson, and Highland Bali: Fieldwork Photgraphs of Bayung Gede, 1936–1939* (Chicago: University of Chicago Press).

Tantri, K. (1960), *Revolt in Paradise* (New York: Harper and Row).

Teeuw, A. (1998), 'Kunjungan Penziarah Sunda ke Bali Sekitar Tahun 1500' (The Visit of a Sundanese Pilgrimage to Bali around the Year 1500), in Aron Mko Mbete (et al.) *Proses dan Protes Budaya* (Denpasar: Balai Bahasa Denpasar dan Bali Post).

Timothy, D.J. and Wall, G. (1997), 'Selling to tourists: Indonesian street vendors', *Annals of Tourism Research*, 24(2): 322–40.

Timothy, D.J. and Wall, G. (1997), 'Selling to tourists: Indonesian street vendors', *Annals of Tourism Research*, 24(2): 322–40.

Toh Mun Heng and Low, L. (1990), 'Economic impact of tourism in Singapore', *Annals of Tourism Research*, 17 (2): 246–69.

Tourism and Travel Intelligence (1998), 'The fall-out from the Asian economic crisis', *Travel and Tourism Analysts*, 6: 78–95.

Udayana University and Francillon, G. (1975), 'Tourism in Bali: its economic and socio-cultural impact', *International Social Science Journal*, XXVIU (4).

van Gemert, H., van Genugten, E. and Dahles, H. (1999), '*Tukang becak*: the pedicab men of Yogyakarta', in H. Dahles and K. Bras (eds), *Tourism and*

Small Entrepreneurs: Development, national Policy and Entrepreneurial Culture: Indonesian Cases (pp. 97–111) (New York: Cognizant Communication Corporation).

Vickers, A. (1984), 'Ritual and representation in nineteenth-century Bali', *Rima*, 18, 1: 1–35.

Vickers, A. (1989), *Bali A Paradise Created* (Singapore: Periplus Editions).

Vickers, A. (1996), *Being Modern in Bali, Image and Change* (pp. 92–114). (Monograph 43/Yale Southeast Asia Studies).

Vickers, A. (1997/8), 'Selling the experience of Bali, 1950–1971', paper presented at Commodification, Seminar Series, 20 August, University of Wollongong.

Vickers, A. (2000), 'Bali Merdeka? Internal migration, tourism and Hindu revivalism. Paper presented at the Society for Balinese Study Conference to Honour Professor Hildred Geertz' (Denpasar, 10–13 July).

Vickers, A. (2003), 'Being modern in Bali after Suharto', in T. Reuter (ed.), *Inequality, Crisis and Social Change in Indonesia, The Muted Worlds of Bali*, pp. 17–29 (London: CurzonRoutledge).

Vickers, A. (et al.) (2000), *To Change Bali, Essays in Honour of I Gusti Ngurah Bagus* (Denpasar: Bali Post and Institute of Social Change and Critical Inquiry, University of Wollongong).

Vickers, A. (ed.) (1994), *Travelling to Bali, Four Hundred Years of Journeys* (Kuala Lumpur: Oxford University Press).

Wall, G. (1991), *Bali Spatial Arrangement Plan: Preliminary Reactions* (Waterloo: University of Waterloo).

Wall, G. (1996), 'Terrorism and tourism: an overview and an Irish example', in A. Pizam and Y. Mansfield (eds), *Tourism, Crime and International Security Issues* (pp. 143–58) (Chichester: John Wiley and Sons).

Wall, G. (1999), 'The role of entrepreneurship in tourism' (un-published conference paper, ATLAS-Asia).

Wall, G. and Long, V. (1996), 'Balinese homestays: an indigenous response to tourism opportunities', in R. Butler and T. Hinch (eds), *Tourism and Indigenous People* (pp. 27–48) (London: Thomson International Business Press).

Wallerstein, I. (1976), *The Modern World System: Capitalist Agriculture and the Origins of the European World Economy in the Sixteenth Century* (New York, Academic Press).

Warren, C. (1991), 'Adat and dinas: village and state in contemporary Bali', in H. Geertz (ed.) *State and Society in Bali, Historical, Textual and Anthropological Approaches* (pp. 213–49) (Netherlands: KITLV).

Wedakarna, A.A.N.A.M.W.S. (2002), 'Mempertanyakan aksi: Bali for the world', *Bali Post* (online), 22 December.

Wijaya, P. (1977), *Tiba-Tiba Malam* (Night Falls Suddenly) (Jakarta: Cyprus).

Wijaya, P. (1995), 'Dasar' (Typical), in J. Lingard (translator), *Diverse Lives* (pp. 93–98) (Kuala Lumpur: Oxford in Asia).

Wilkinson, P.F. and Pratiwi, W. (1995), 'Gender and tourism in an Indonesian village', *Annals of Tourism Research*, 22(2): 283–299.

Williams, L. and Putra, D. (1997), 'Cultural tourism: the balancing act', discussion paper (New Zealand: Lincoln University).

Wirawan, A.A.B. (1995), 'Puputan sebagai swadharmaning negara, persepsi sejarah' (*Puputan* as devotion to the state, an historical perspective), in I.W. Supartha (ed.), *Dharma Agama and Dharma Negara* (The devotion to religion and state) (pp. 95–112) (Denpasar: PT Bali Post).

Wolf, E. (1957), 'Closed Corporate Communities in Mesoamerica and Central Java', *Southwestern Journal of Anthropology*, 13: 1–18.

Wollen, P. (1993), *Raiding the Icebox, Reflections on Twentieth Century Culture* (London: Verso).

Wood, R.E. (1993), 'Tourism, culture and the sociology of development', in M. Hitchcock, V.T. King, and M.J.G. Parnwell (eds), *Tourism in South-East Asia* (pp. 48–70) (London: Routledge).

Woodward, M. (1989), *Islam in Java: Normative Piety and Mysticism in the Sultanate of Yogyakarta* (Tucson: The University of Arizona Press).

Yamashita, S. (2003), *Bali and Beyond: Explorations in the Anthropology of Tourism* (New York: Berghahn Books) (Translation and Introduction by J.S. Eades).

Newspaper Articles

Bali Update, 2006, 'Illegal Levies Destroying Tourism's Image', 21 May 2006.

Sarad, 2002, Mengincar kepala gumi Bali, *Sarad* 22, January 2002, 28–31.

Nusa, 2001, Tim sosialisai warisan budaya dunia dibubarkan, August 14.

Kompas, 1993, Mendikbud Fuad Hassan: Pura Besakih tak akan dijadikan cagar budaya, January 12.

Denpost, 2001, Taman Ayun dijajaki jadi warisan budaya dunia, August 10.

Bali Post, 1990, Gubernur: Pura Besakih tidak perlu dijadikan warisan dunia, October 4.

Bali Post, 2001, Menbudpar Gde Ardika: Usulkan Besakih jadi warisan budaya dunia, October 1.

Bali Post, 2001, Ardika dituntut mundur, October 4.

Bali Post, 2001, Menneg budpar Ardika: Jadi WBD Pura Besakih tak berubah status, October 5.

Bali Post, 2001, Putra Agung: Tak ada masalah, October 6.

Bali Post, 2001, Besakih tetap diajukan jadi warisan budaya dunia, December 15.

Bali Post, 2001, Ardika tunda usulan Pura Besakih jadi WBD, December 23.

Bali Post, 2001, Jika dijadikan WBD, Besakih akan banyak dikunjungi wisman, December 24.

Bali Post, 2002, Diprotes, 'charge' Rp 30.000 di Besakih, 11 May.

Contours, Supplement to vol. 7, no. 7 (September 1996).

Bali Post, 2003a, Tingkat hunian rendah, hotel banyak dijual dan jadi rumah kost (Occupancy low, many hotel are sold and put up for rent), 9 August, 10.

Bali Post, 2003b, Pelabuhan Gilimanuk akan dilengkapi eksplosif detektor (Gilimanuk Harbour to be equipped with explosives detector), 9 August, 3.

Bali Post, 2003c, Kapolda akan keluarkan standar pengamanan hotel (Bali's regional police will issue standards of hotel security), 14 August, 2.

Bali Rebound, July 10–July 24, 2004, Jakarta: *The Jakarta Post*.

Kompas, 2006, Pariwisata Bali Sekarat! (Bali's Tourism in Agony), Wednesday, 11 January, 34–35.

Telegraph Travel, 13 June 1998, 'Bali: We're Safe', A. Spillius.

Travel Asia, 22 May, 1998.

TNT UK, 3 August, 1998.

Down to Earth, 23 April 1994.

The Advertiser, 29 April 2006, 'How Bali bombers planned mission', C. Wockner.

The Bali Times, 15–21 July 2005.

The Daily Telegraph, 26 April 2006, 'Bali bomb code', C. Wockner.

The Economist, 26 July 1997.

Northern Region News, April 3 2006.

Sydney Morning Herald, 3 December 1999.

Radio Australia AM Report, 2006, 'Report questions security in Indonesia', Saturday 6 May, 2006, 08: 15: 00.

Internet-based References

Berry, E.N. (2001), 'An application of Butler's (1980) tourist area life cycle theory to the Cairns region, Australia, 1876–1998' (PhD Thesis, Tropical Environment and Geography, James Cook University, Cairns Campus). www.geocities.com/tedberry_aus/tourismarealifecycle.html.

Index